Lecture Notes in Chemistry

Volume 86

The Lecture Notes in Chemistry

The series Lecture Notes in Chemistry (LNC) reports new developments in chemistry and molecular science-quickly and informally, but with a high quality and the explicit aim to summarize and communicate current knowledge for teaching and training purposes. Books published in this series are conceived as bridging material between advanced graduate textbooks and the forefront of research. They will serve the following purposes:

- provide an accessible introduction to the field to postgraduate students and nonspecialist researchers from related areas,
- provide a source of advanced teaching material for specialized seminars, courses and schools, and
- be readily accessible in print and online.

The series covers all established fields of chemistry such as analytical chemistry, organic chemistry, inorganic chemistry, physical chemistry including electrochemistry, theoretical and computational chemistry, industrial chemistry, and catalysis. It is also a particularly suitable forum for volumes addressing the interfaces of chemistry with other disciplines, such as biology, medicine, physics, engineering, materials science including polymer and nanoscience, or earth and environmental science.

Both authored and edited volumes will be considered for publication. Edited volumes should however consist of a very limited number of contributions only. Proceedings will not be considered for LNC.

The year 2010 marks the relaunch of LNC.

More information about this series at http://www.springer.com/series/632

Roman Pampuch

An Introduction to Ceramics

Roman Pampuch
University of Science and Technology
Kraków
Poland

The Work was first published in 2013 by Wydawnictwa AGH with the following title: Wykłady o Ceramice.

ISSN 0342-4901 ISSN 2192-6603 (electronic)
ISBN 978-3-319-36350-9 ISBN 978-3-319-10410-2 (eBook)
DOI 10.1007/978-3-319-10410-2

Springer Cham Heidelberg New York Dordrecht London

Printed on acid-free paper

Springer is part of Springer Science+Business Media (www.springer.com)

Preface

Ceramic materials are usually defined by specifying what they are not. They are called inorganic, because they do not consist of electrically neutral molecules—which are typical for organic compounds. They are also called non-metallic, because ceramics are characterised by a gap between allowed energy bands of valence electrons, a feature not existing in metallic materials.

Applications of ceramic materials are primarily determined by their inherent properties. Traditional ceramics take the advantage of immanent properties of minerals and rocks of the earth's crust. Clays, feldspars and quartz—are used to produce table, sanitary and other kinds of traditional ceramics, clays rich in limestone and quartz sand (margles) to make Portland cement clinker. Typical glass, the soda-lime-silica glass, is made by melting lime-, quartz- and soda-bearing raw materials. Natural magnesite, dolomite, fireclay are among rocks exploited to fabricate refractories. The range of properties has been widened by the advent of advanced ceramic materials, produced from synthetic compounds and deeply transformed natural raw materials. The advanced ceramic materials can be classified, according to the state of their valence electrons and chemical bond type, into four groups having similar immanent properties. Namely: (i). covalent semiconductors (including a.o. Si, SiC and GaP); (ii). Ionic semiconductors (including a.o. CdS, GaN and GaAs); (iii). ionic dielectrics (including a.o. $BaTiO_3$, $Pb(Zr_yTi_{1-y})O_3$ and $Pb(Mg_{0,33}Nb_{0.66})O_3)$; (iv). covalent dielectrics (including a.o. Al_2O_3, ZrO_2, Si_3N_4, and diamond). Exploitation of advanced ceramic materials is often feasible if the materials acquire new useful properties by adequate processing.

The title "Introduction to Ceramics" indicates that the present book is intended not to be another of the numerous textbooks or compendia about ceramics but should complement them. In order to help the reader grasp the main points, it is written in the form of concise essays. A book like this seemed to be timely in view of the rising flood of information and the reigning lack of time for reflection, needed to convert information into knowledge. However, this is essential for making the right choices and acting effectively in a dynamic world we are living in. Groups for which such a book may be useful include senior undergraduates choosing studies direction, students of physics and chemistry who have to choose a specialization;

already specialized doctorial students and engineers to gain a broader perspective, and businessmen, because It is of key importance to make business-related decisions on the basis of a broad overview of issues. The broad range of potential readers require the use of plain language and the simplest explanations possible, while a knowledge of basic notions of physics and chemistry could be assumed here.

Kraków, September 2013 Roman Pampuch

Contents

Chapter 1
A Brief History of Ceramic Innovation

1.1 Innovation Resulting from Natural Knowledge

Neolithic Revolution Innovations in the field of ceramic materials has played an important role in all critical epochs, called revolutions by sociologists and historians (Fig. 1.1). The first of these was the Neolithic Revolution. In the Middle East, its origins date back to around 10,000 years. On one hand, it was an agricultural revolution associated with the transition from a nomadic gatherer and hunter lifestyle to a settled lifestyle. This made it possible for farmers to produce an excess of food, and enabled the development of breeding livestock. On the other hand, it was an urban revolution, which was manifested in the formation of large societies in Mesopotamia, Egypt, Indus valley, China and Central America, concentrated in small areas, i.e. cities, where hardly anyone worked in agriculture. Consequently, division of labour emerged, which favoured the multiplication of material and spiritual goods.

Early Use of Clay An important element of the Neolithic Revolution in some regions was the use of clay. Clay is sedimentary rock, made of clay minerals, quartz gravel captured during sedimentation and organic components. This rock, typical for river valleys, became, according to archaeological evidence, the basic structural material in Mesopotamia and in the Indus Valley. The key in this case was surely the discovery that clay, when mixed with a specified amount of water, becomes easy to form under pressure into different shapes which are retained when the pressure is relieved. Bricks of dried clay, sometimes reinforced with chaff, are still used for building houses in the Middle East. According to archaeologists, the move to the next stage, i.e. firing of shaped clay–water mixture, dates back to around 5,000 years, and was probably favoured by the discovery that, at temperatures achievable when burning wood (1,000–1,100 °C), the dried clay–water, mass turns into a relatively tough body (→ 2. Tradition continued. Clay tradition). At least the external cladding of temple hills (ziggurats) in Mesopotamia and the walls of other important structures—as Herodotus states in *The Histories*—were made of more durable fired bricks. The tradition of using fired clay bricks has survived until the present day.

R. Pampuch, *An Introduction to Ceramics*,
Lecture Notes in Chemistry 86, DOI 10.1007/978-3-319-10410-2_1

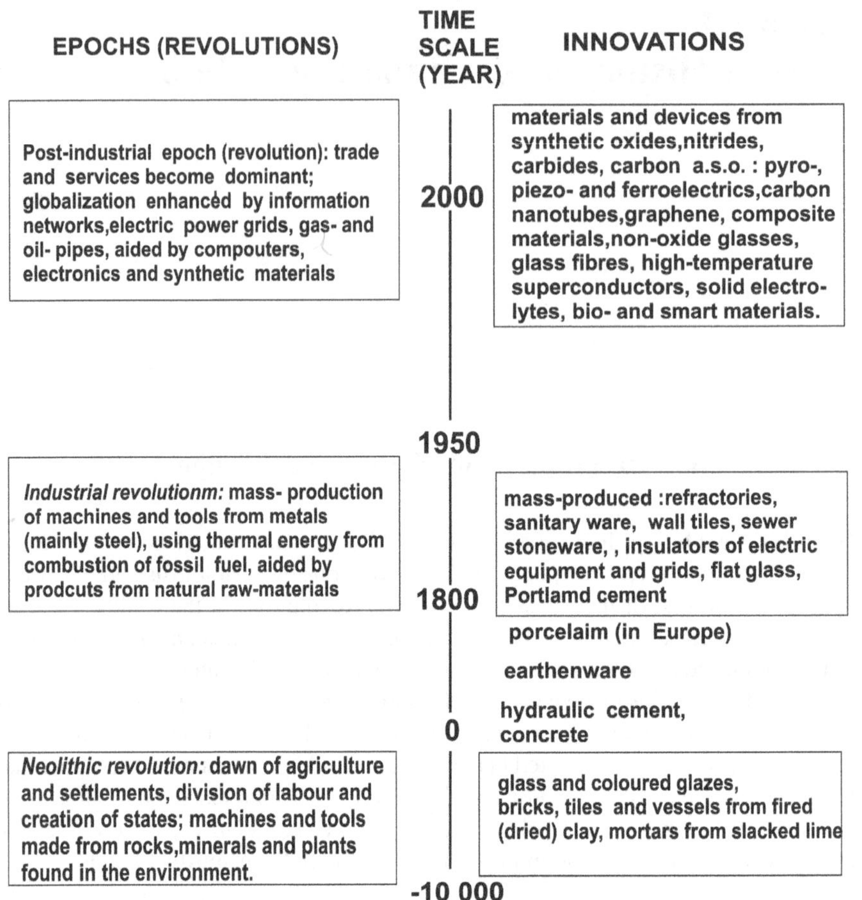

Fig. 1.1 Successive epochs of civilisation (in Europe and the Near East) and major innovations in ceramics

An equally important role in the Neolithic Revolution was played by at least two more innovations involving the use of clay. The first was the advent of impervious and tough pottery made by firing the formed clay–water mass. This pottery was suitable for the central storage and distribution of food, which was necessary in the face of labour division and territorially organised societies. Such vessels had been impractical for the previous nomadic life, in which frequent packing and transport of hard but brittle objects would have caused a lot of trouble. The transition from forming the clay–water mass by kneading to shaping it by turning on a potter's wheel probably enabled the production of clay pottery to be become more 'industrial'.

Another innovation was the invention in southern Mesopotamia (Sumer), around 5,000 years ago, of writing on clay tablets, prisms and rolls, many of which have survived thanks to being fired. In addition to texts devoted to religion, morality, literature and social philosophy, records of trade transactions and legal acts predominate. This was indispensable for efficient political administration and long-distance trading. Such records are, historically, the first example of information storage and its transmission to future generations.

Slurry Casting in the Hellenistic Era In the following centuries, the art of making ceramic pottery of formed and fired clay-water mass attained a high level in Greece, as early as in the times of Homer. From the time of Pericles (2,500 years ago) come the short ceramic pipes used for water supply in Athens. A breakthrough in ceramic technology was the emergence, in the Hellenistic period (2,000–2,300 years ago), of products formed by casting aqueous clay suspensions (slurries) in moulds. The clay-water slurry easily adapts even to complex geometry within a mould and, after most water is sucked out of the mass, cohesion of deposited solid particles increases enough to allow further processing. To create relatively stable slurry, the agglomeration of solid particles dispersed in water must be prevented. In Hellenistic slurries, this was achieved—according to assorted evidence—either by using potassium-rich clay or by adding alkali-rich wood ash. As is known nowadays, the presence of a low concentration of K^+ or Na^+ ions in aqueous solution leads to their sorption on the surface of solid particles and stabilisation of the suspension (→ 2. Tradition continued. Clay tradition).

Earthenware During the Renaissance, earthenware became widespread in Europe. It was a glazed kind of earthenware known from excavations from the Indus valley and ancient Egypt and Minoan Crete. These were products with a body made by forming and firing mixtures comprising mainly a clayey ingredient, quartz sand and feldspars. The clayey ingredient was marl, a clay containing significant amounts of fine calcium carbonate. The earthenware body in majolicas was coated with an amorphous and transparent glaze (→ 2. Tradition continued. Silica tradition and 6. Materials versus light) appearing in products from around 3,000 years ago, found in Mesopotamia. The glaze was most probably obtained by application on a fired body of a mixture, containing mainly clay with quartz grains, rich in sodium vegetable ash (of a Mediterranean plant, *salsola soda*) and calcium carbonate. The entire product must have been then fired again. Evidence indicates that in the ninth century, in Persia, a method of introducing a 4–10 % volume of so-called opacifiers was adopted from China. These opacifiers were particles of tin dioxide (SnO_2) insoluble in molten silica of the glaze. When exposed to light, the particles caused diffuse reflection of light (→ 6 Materials versus light). By way of Mauritania and Majorca (from which originates the name *majolica),* the product reached Europe. Its production was developed mainly in Faenza in Italy, from which the name 'faience' is derived. A typical innovation introduced in Faenza was coloured enamel glazing.

Porcelain Porcelain tableware and works of art have been manufactured in Europe since the eighteenth century. Porcelain is a noble ceramic material with a dense white body, impermeable to water and gases. Thin-walled porcelain allows the passage of light. For the production of porcelain, a process similar to that used for earthenware was adopted, but less impure and finer ground mineral raw materials were used. The main raw materials used here included kaolin, a clay rich in the clay mineral: kaolinite, pulverised quartz and feldspars (sodium, potassium and calcium and, more rarely, barium aluminosilicates). Obtaining porcelain from such raw materials required the application of higher firing temperatures than before. The critical factor behind the start of porcelain production in Meissen (Saxony) in 1708 was probably von Tschirnhaus's invention of suitable furnaces which enabled the attainment of high temperatures. The name of porcelain was disseminated in French, in which the word 'porcelaine' (from Latin *porsella*) referred to the pearl mass the porcelain was believed to resemble. The earliest porcelain products, however, first appeared in China, a early as 1,800–1,900 years ago. The traditional English name 'china' refers to this fact.

Early Oxide Glass In a broad sense, classical ceramic materials made of natural raw materials include glass (→ 2. Tradition continued. The silica tradition). There is evidence that as early as in ancient times it was discovered that a mixture of solid components (containing around 65 % SiO_2 and 30 % Na_2O + CaO, as we know today) forms a liquid at not very high temperatures and, upon cooling, solidifies, forming a transparent material: glass. From these and similar materials, the manufacture of glass products was started in ancient Phoenicia and in Egypt. Ingots were probably made originally by melting ground quartz sand, clay and sodium-rich plant ash in ceramic vessels. After cooling, the ingots were then presumably heated again and formed into the desired shapes. In the Roman Empire, the technology of pouring molten glass-forming compositions into moulds was developed. This enabled mass production and availability of glass vessels within broader social circles than before. The Middle Ages in Europe are characterised by the development of stained-glass windows and, since the twentieth century, glass has been widely used in construction.

Of similar extent is the history of cements. Composed of clay, gypsum and sand, the mortars for permanent bonding of stone or brick were already being used in ancient Egypt, and quicklime mortars, as far as we know, appeared first in Crete and, on a larger scale, in the Roman Empire. This was thanks to an early discovery of the specific properties of some natural raw materials (→ 2. Tradition continued. Lime tradition) of which we know now that that they contain Ca–O groups and calcium aluminosilicates; and are highly reactive towards air and water. The first cement setting (hardening) on reaction with water, i.e. a hydraulic cement, mentioned historically, was made in Roman Empire. It was prepared from a mixture of quicklime (made by firing chalk and seashells) and volcanic ash from the Pozzuoli region. This ash contained the highly reactive calcium silicates and aluminosilicates. Such materials are usually called *pozzolanic* after the name of the site. Pozzolanic hydraulic cement, after blending with water, was used for bonding aggregate and sand mixed with it. A hard and strong material, namely concrete, was formed in this way. The Pantheon in Rome that has survived to date, was built in the first century BC using, among others, structural blocks made of this concrete.

1.2 Innovation Based on Geological and Mineralogical Knowledge

Ceramic Products of the Industrial Revolution Another turning point in civilisation was the Industrial Revolution, which began around 200 years ago in England. One of its distinct features was the extensive combustion of fossil fuels (coal, and later gas and oil) for the generation of heat and steam (and electricity). Another indicator was the mass production of steel and steel tools for the creation in turn of more machines and tools, which thus became widely available. A fundamental role in the Industrial Revolution was played by metals, especially steel.

However, to meet the increasing demands, boosted by the Industrial Revolution, of industry and society, mass production of ceramic products from natural raw materials began. Depending on the type of product, various proportions of clay-, quartz- and feldspars were used, and methods proven in the production of earthenware and porcelain applied (→ 2. Tradition continued. Clay tradition)). The products include: a. big high-voltage insulators for power supply lines, made in most cases of so-called technical porcelain which contains more quartz and Al_2O_3 than traditional porcelain; b. sanitary ware—this term includes a series of products used for sanitary purposes, such as washbasins, toilet bowls, bidets, etc.; c. Acid-resistant ceramics used for drains and sewers; d. Ceramic tiles for walls and floors.

Refractories This was accompanied by the parallel development of industrial production of refractory lining and heat insulation materials applied in furnaces for melting iron, steel, non-ferrous metals and glass, firing lime and making coke from coal. Products derived from refractory clay with an increased Al_2O_3 content, were later supplemented or replaced by refractory materials made of natural oxide minerals, having high melting points and resistance to high-temperature corrosion (→ 2. Tradition continued. Refractory tradition).

Portland Cement The most important innovation in the field of ceramics during the Industrial Revolution was, however, hydraulic Portland cement (→ 2. Tradition continued. Lime tradition). Experimental science, evolving as of the eighteenth century, indicated that all materials containing calcium react readily with water, and, to obtain hydration products with good mechanical properties, it is necessary to use raw materials in which calcium and silicates occur together. In 1824, Aspdin patented a method for the production of so-called Portland cement clinker by firing marl, i.e. clay rich in limestone and quartz sand, at 1,400–1,450 °C. Today, we know that when these raw materials are fired, reactions occur which lead to the formation of calcium silicates according to the general formula: $5CaCO_3 + 2SiO_2 = (3CaO\bullet SiO_2, 2CaO\bullet SiO_2) + 5CO_2$.

Portland cement and concrete made from it, various kinds of oxide glass, products made from clay–feldspar–quartz mixtures and refractories, have formed the universal image of ceramics to this day (Table 1.1).

Table 1.1 Estimated global production of steel and main ceramic products

	Global production (10^9 tonnes/year)	Production per adult on earth (kg/year)
Steel	3.5	~300
Ceramic tile	0.005	~1
Refractory materials	0.04	~8
Flat glass products	0.006	~1
Portland cement	3.5	~300

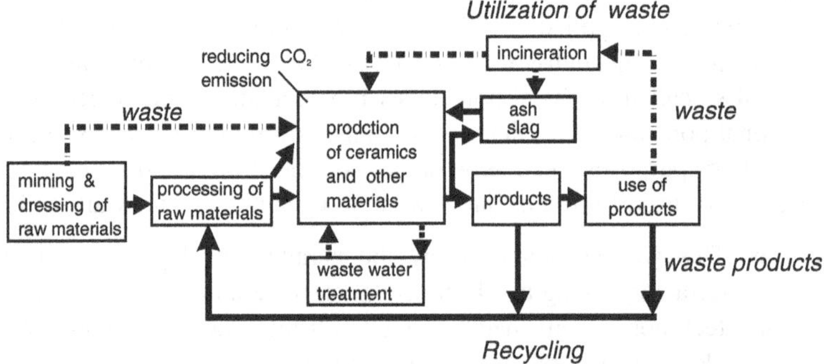

Fig. 1.2 The modern industrial ecosystem

Industrial Ecosystem As a result of globalisation, at the end of the twentieth century, the Industrial Revolution system began to approach limits which, if exceeded, might cause irreversible harm to human existence on the earth. In traditional ceramic technologies, this resulted in the formation of a system promoting reasonable management of the earth's resources (Fig. 1.2). For instance, the use of non-combustible waste clay (gangue), a by-product of mining for such fossil fuels as hard coal and lignite, for ceramic production. Furthermore, the utilisation of slag, a by-product of smelting metals from ores and purification (refining) processes, containing mainly calcium and magnesium silicates. Given their similar phase composition and properties, slag, together with Portland cement clinker, can be used for the production of hybrid hydraulic cements. Other traditional ceramic 1 products such as bricks (from fired clay) tolerate—without impairment and, sometimes, with improvement of their properties—large volumes of waste, such as volatile ash from burning coal and oil. Since, volatile ash and slag contain SiO_2 and Al_2O_3, they can be used for the production of a specific group of binding materials: geopolymers (→ 2. Tradition continued. Lime tradition).

Incineration and Immobilisation of Heavy Metals One of the typical characteristics of the Industrial Revolution era is growing heaps of waste. These are best disposed of by incineration, which, however, requires quality refractory linings in the incinerators. This is particularly true when incineration temperatures are increased to around 1,400 °C, so as to eliminate the complex system of filters required when burning products containing harmful heavy metals. There is also a problem of immobilisation of such heavy metals as Pb, Cu, Cd, Zn, Ag in the disposal of sludge (sewage) in water treatment plants. In addition to organic compounds, sludge contains aluminosilicates which can be used for making mortar and construction materials, using mixtures of sludge and binding material such as slag. The process of transformation of sludge and sewage into construction material enables the immobilisation of the harmful heavy metal ions they contain. An even greater challenge is the immobilisation of radioactive isotopes with long half-lives present in nuclear waste. Boron-containing glass, with a large active cross-sectional for nuclear radiation, is broadly used to hold the nuclear waste and radiation within a 'sarcophagus'.

Reduction of Gas Emission Traditional ceramic materials are also used in equipment reducing the emission of toxic gases as NO_x or greenhouse gases (CO_2). Such gases are generated on a large scale by combustion engines. To prevent this, converters are commonly used in vehicle exhaust systems. On the one hand, they serve as filters, by retaining solid particles suspended in the combustion gas. On the other hand, owing to their considerable surface area, they serve as substrates for catalysts of reduction reactions of NO_x—N_2 and oxidation of CO and hydrocarbons to CO_2 and H_2O. Since there are extreme temperature variations in these exhaust systems and the combustion gases are often aggressive, the converter material must be resistant to high-temperature erosion and corrosion and to thermal shocks (rapid temperature changes (→ 2. Tradition continued. Refractory tradition)). All these requirements are well met by cellular honeycomb ceramic materials (Fig. 1.3) obtained by forming plastic masses of cordierite (magnesium aluminosilicate). Ceramic filters also play an important role in removing inorganic solids contained in liquid metals.

Rational Water Management Since, the early days of the Industrial Revolution, an important contribution to rational water management has been made by sanitary wares finished with high- quality glaze (→ 2. Tradition continued. Clay tradition; 6. Materials vs. light) for easier cleaning and by *acid-resistant ceramics*. Later on, this contribution was expanded by the application of membranes and ceramic filters for filtering or clarifying such products as beer, wine and other beverages or skimmed milk (→ 4. Voids are important) and the use of ceramic compounds for ion exchange. In the latter case, synthetic oxides and, even more often, aluminosilicates of natural origin are used. These are: dehydrated zeolites or clay minerals from the smectite group (montmorillonites), containing low soluble Na^+ and K^+ cations which are readily exchanged with Ca^{2+} and Mg^{2+} ions present in 'hard' water.

Degradation of Water Pollutants by Photocatalysis Rational water management involves a disposal and treatment of waste and polluted water. For the disposal of it outside the habitations pipes and tubes, made first of fired clay and then of acid

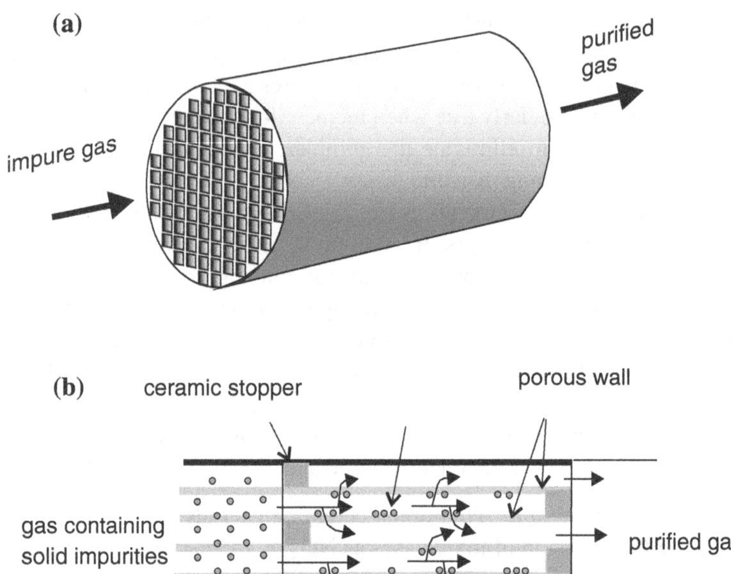

Fig. 1.3 Catalytic converter made of cordierite (a) and a diagram of its operating principle (b)

resistant clayey ceramics, have been used. The increasing pollution of utility water with organic polymers, dyes, surfactants, pesticides and bacteria requires the degradation of such pollutants to nontoxic compounds such as CO_2 and H_2O. Since the turn of the twenty-first century, the unique photocatalytic properties of titanium oxide have been used for that purpose. In TiO_2, under the influence of photons of light with energy *hv,* the reaction $TiO_2 + hv \rightarrow TiO_2 + (e' + h\bullet)$ occurs, in which free electrons, e′-and electron holes h• are generated. These are capable of oxidising or directly reducing bacteria or organic water pollutants, by either removing or adding electrons. Oxidation and reduction may also convert water or oxygen molecules into radicals, i.e. molecules containing non-paired electrons, as in the following reaction: $h\bullet + H_2O \rightarrow H_2O^\circ$; $e' + O_2(ads) \rightarrow O_2^\circ$. The symbol (°) denotes a non-paired electron. The non-paired electrons of the radicals remove electrons from polymers, dyes, surfactants and pesticides, thus causing their oxidation and degradation.

1.3 Innovation Based on Chemistry, Physics and Technical Sciences

Post-industrial Civilisation The second half of the twentieth century saw important changes which are considered by many as the beginning of a new, post-industrial civilisation, sometimes referred to as the information revolution. From the economic point of view, this civilisation is characterised by the increased importance of trade and services. From the technological perspective, it is identified with the widespread use of electronics and computers linked by information technology (IT) systems which exchange information at the speed of light. There is also the spread of power systems, oil and gas pipelines in order to ensure sustainable supply and demand on at least a continental scale. To meet the challenges of this civilisation, materials science and engineering was created, an interdisciplinary scientific domain which aims mainly at utilising scientific knowledge for the effective and economic design and manufacture of metallic, polymer and ceramic materials.

Synthetic Powders, Layers and Monocrystals Advent of materials science and engineering has resulted in an increased demand for the unique properties of ceramic materials based on synthetic inorganic, non-metallic compounds as raw materials (Table 1.2). To achieve this, many new and improved methods of synthesis and generation of products were required. These are mainly methods utilising shaping and sintering of synthetic ceramic powders and their suspensions, often colloidal ones, chemical and physical layer deposition from the vapour phase, crystallisation from liquids, thermal decomposition of organic compounds (Figs. 1.4, 1.5 and Table 1.3) and many other methods.

Ceramic Structural Materials: Replacing Metals An important category of advanced ceramic materials comprise structural materials, developed, still in the spirit of the Industrial Revolution, to replace metals in all applications where higher rigidity, hardness and resistance to wear and an aggressive environment is required

Table 1.2 Comparison of typical properties of ceramics, metals and polymers

Property	Metals	Polymers	Ceramics
Tensile strength (MPa)	100–1,500	1–100	100–900
Compressive strength (MPa)	100–1,500		1,000–5,000
Fracture toughness (MPa $m^{-1/2}$)	10–30	2–8	1–10
Relative strain at failure (%)	4–40	2 (1,000 elastomers)	1
Density (g cm^{-3})	2–20	1–2.5	1–14
Thermal conductivity (W m^{-1} s^{-1})	50–350	Low	2–100 (graphite and nanotubes 600–3,000)
Thermal expansion ($\times$ 10^{-6} K^{-1})	7–19	100	2–11
Max. Application temperature (°C)	900	250	1,400–1,720
Resistance to high-temperature corrosion	Low/ medium	Low	High
Electrical conductivity	High	Low	Low
Transparency in visible light	Non-transparent	Non-transparent /transparent	Non-transparent (semiconductors)/transparent (dielectrics)

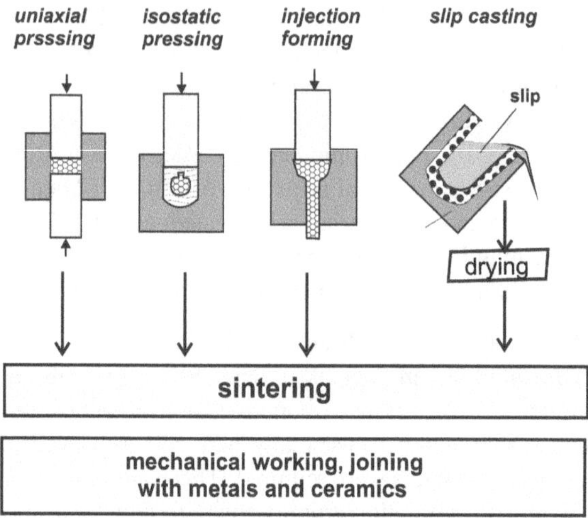

Fig. 1.4 Schematic presentations of basic methods of shaping and fabricating ceramics. Note: In the case of extrusion, homogeneous mixtures of ceramic powder and polymer are most frequently used, and extrusion is performed at a low or increased temperature under high pressure; in the case of casting in moulds, a broad range of methods have been developed, from shaping of a rigid product by sucking liquid out of the deposit in the mould, to direct deposition on the mould surface of colloidal particles and of gels created in a suspension of colloidal particles (sol)

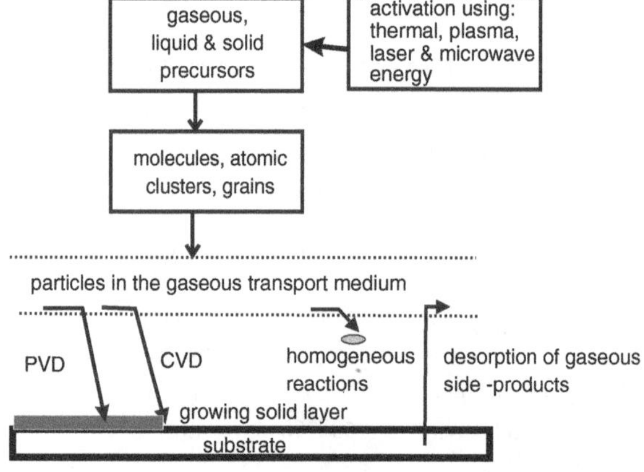

Fig. 1.5 Main stages of physical (PVD) and chemical vapour deposition (CVD) on solid substrates

(Table 1.4). From that perspective, attention was drawn to αAl_2O_3, SiC, Si_3N_4, or sialons (Si_3N_4, in which part of the Si is replaced by Al, and N by O). These compounds are composed of elements located near one another in the periodic table of elements and thus have similar effective atom nuclei charges Z_{ef}. Due to this bond in these compounds have a high covalent contribution (→ 5. Unusual dielectrics and conductors) and form a three-dimensional net, distinguished by great rigidity, hardness and wear resistance (→ 3. Ceramics to overcome brittleness).

Table 1.3 Some advanced methods for shaping and for fabrication of ceramic materials, other than powder compaction and casting of slurries

Raw material	Shaping method	Shape of the productial
Bed of loose powder	Forming, layer by layer, of agglomerations of particles and binder sprayed on the particle bed in a controlled way	3D[a]
Powder suspension in photosensitive organic monomer	Forming, layer by layer, of powder-polymer agglomerations of a shape determined by exposure to laser beam	3D[a]
Water suspension of powder and water-soluble organic monomer	Gel casting: casting in moulds followed by monomer polymerisation	3D[a]
Colloidal suspension of inorganic–organic precursor (e.g. alcoholate)	Sol–gel method: coagulation and polymerisation with formation of a gel which can be applied on a substrate	2D[a] on a substrate[a]
Powder suspension	Screen printing: application on a substrate through holes in a mask (fabric, metallic mesh)	2D[a] on a substrate
Mixture of powder and polymer	Application of the mixture on a substrate	Organic foil containing powder particles
Volatile molecules, atoms or clusters of atoms	Molecular beam deposition of epitaxial layers on a substrate, chemical or physical vapour deposition of layers on a substrate	2D on a substrate
Continuous polymer fibres	Polymer pyrolysis	Continuous ceramic fibres

Notes [a] 2D—two-dimensional; 3D—three-dimensional shape

From such materials are made, among others, high-wear parts of structures such as bearings, mechanical closures and sealing rings, pulveriser and welding nozzles, grinding media and mill linings. Polycrystalline SiC, Si_3N_4, ZrO_2 and αAl_2O_3 are used for manufacturing full-ceramic or hybrid rolling bearings, particularly for water pumps where, in addition to the mentioned properties, high resistance to corrosion is important. The application of polycrystalline αAl_2O_3, Si_3N_4 and diamond as tools enables metal machining which is faster by two orders of magnitude compared to fast-cutting steel (Fig. 1.6). The properties of this group of ceramic materials have also been used in making bulletproof vests and armour plates for attack vehicles. In the period immediately following the introduction and improvement of high-explosive anti-tank warheads, no metallic armour could resist them. This has changed since SiC, B_4C or αAl_2O_3 armours have been in use. These lightweight materials—hard and brittle but high-melting—when smashed into pieces, distort the geometry of a stream of very hot metal occurring at an anti-tank warhead strike, thus preventing further penetration into the armour. To limit damage to the place of impact and for technical reasons, ceramic armours consist of small segments which can be easily replaced by bonding them to the metal structure of a vehicle.

Table 1.4 Some applications of construction ceramics made of synthetic raw materials

Property of particular interest	Main applications	Typical material
High hardness	Mill linings, grinding agents	Al_2O_3
High hardness and abrasion resistance	Pulveriser nozzles, welding nozzles, mould linings, press tools and spinning nozzles	Al_2O_3, Si_3N_4
High hardness, brittleness and thermal resistance	Armour for attack vehicles, plates for bulletproof vests	Al_2O_3, SiC, B_4C
High hardness, abrasion resistance and thermal stability	Machining tools for metals and abrasives	Al_2O_3, Si_3N_4, diamond, cBN, SiC
High hardness, abrasion resistance, thermal and chemical stability	Sealing rings, hybrid ball bearings operating in aggressive environment and in automotive industry	Si_3N_4, SiC, Al_2O_3
High-thermal stability	Mufflers and furnace chambers	Al_2O_3
High-thermal stability and low-specific gravity	Turbocompressor rotors	Si_3N_4
High-thermal stability and thermal conductivity	High-temperature heat exchangers, recuperative burners	SiC
Thermal stability and low reactivity with metals	Parts of equipment for transport of liquid metals (e.g. stopper rods, tundish stoppers and ladles) and for hot metals (setters, rollers)	Al_2O_3, SiC, ZrO_2
High resistivity, low-thermal conductivity, resistance to thermal shocks and low reactivity to metals	Induction, dielectric and microwave heaters in contact with metals	Si_3N_4, SiC, MgO

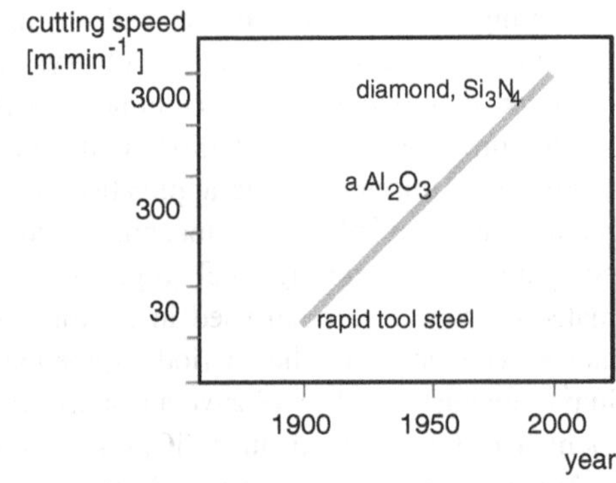

Fig. 1.6 Cutting speed in metal working over the years. *Note* typical tool materials enabling a given cutting speed are specified

Polymer Matrix Composites Where low-specific gravity is important, as in the case of aviation construction, an important role in supplementing and even replacing metals has been played by polymer matrix composites reinforced with continuous ceramic fibres, mostly carbon and glass fibres (→Figs. 3.12. Tradition continued. Silica tradition, respectively). The graphite-like structures of carbon fibres are virtually free from defects and offer strong C–C bonds. Therefore, the tensile strength of carbon fibres may reach several GPa, with high rigidity characterised by a Young's modulus up to 1,000 GPa. Such composites are used for the production of various types of sports equipment and prosthetic limbs; one spectacular example is the manufacture of most structural parts of the Boeing Dreamliner jet airliners.

Micro Reactors and Heat Exchangers The domains in which ceramic materials can successfully replace metals also include heat exchangers and micro-reactors for chemical reactions. In the micro-reactors and micro-heat exchangers cylindrical pores are formed with diameters measured in microns. This results in a high ratio of surface area and volume of liquid flowing through them, which enables efficient exchange of heat with the surroundings. The high surface to volume ratio permits also to realise highly exothermic reactions. A high velocity of a laminar flow of the at a low pressure gradient is typical for channels of a diameter and this enables an output of chemical reactions between fluid reactants of tens of thousand tonnes per annum With dimensions as small as a postage stamp, micro-reactors can be made of SiC, Al_2O_3 or glass to effectively exploit their high chemical and thermal resistance (→ Fig. 4.4).

Ceramics for Combustion Engines Combustion engines are still only a potential area of application for SiC, Si_3N_4, αAl_2O_3 or MgO. Up to temperatures of 1,500 °C, such materials sustain a high rigidity and mechanical strength as well as a low creep rate (ca $1 \cdot 10^{-8} s^{-1}$) allowed for jet engines. Using such materials, the temperature of jet engines could be increased by around 500 °C in comparison to the use of the best metals, such as Ni-based superalloys, thus providing higher efficiency and reduced emission of harmful gases (NO_x and CO). However, brittleness is still a barrier which has not been yet fully overcome (→ 3. Ceramics to overcome brittleness).

Semiconductor Devices Semiconductors and dielectrics making up the advanced ceramic materials have in common the presence of an energy band gap between the ground and the first excited state of valence electrons. This gives rise to various properties which are essential for the post-industrial civilisation. In thin layered semiconductors of the size of a fingernail, electronic integrated circuits are created mainly by lithography (Fig. 1.7), with a number of interconnected active electronic components, such as transistors and diodes, reaching ten million per square millimetre. Semiconducting diodes and transistors are characterised by an exceptional ability to effectively switch between 'current flow' and 'no current flow' states. In the mathematic binary system used in computers and other digital equipment, such states correspond respectively to the numbers (bits) '1' and '0' (Fig. 1.8). Such properties have led to the widespread use of integrated circuits in computer memory and microprocessors. Microprocessors form the central processing units of computers and are used in many other areas, from everyday items, cars and mobile phones to applications in industrial process control.

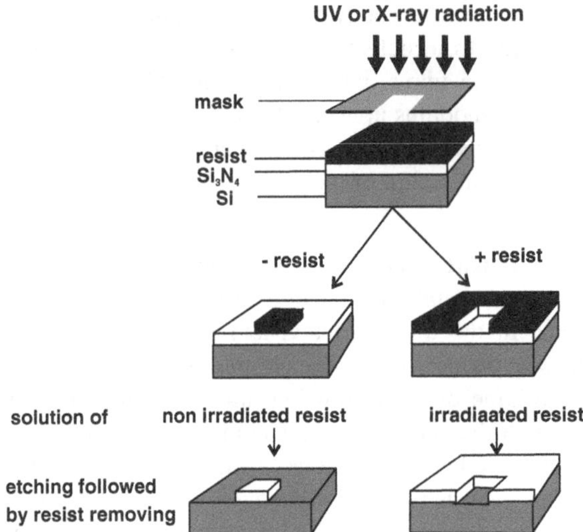

Fig. 1.7 Successive operations in lithography. *Note* +resist is a photosensitive polymer which, after exposure to light, becomes soluble in a developer, while unexposed parts remain insoluble. −resist, becomes cross-linked during exposure, which reduces its solubility; only unexposed parts can be dissolved. Typical developers are alkaline aqueous solutions of organic amines or KOH

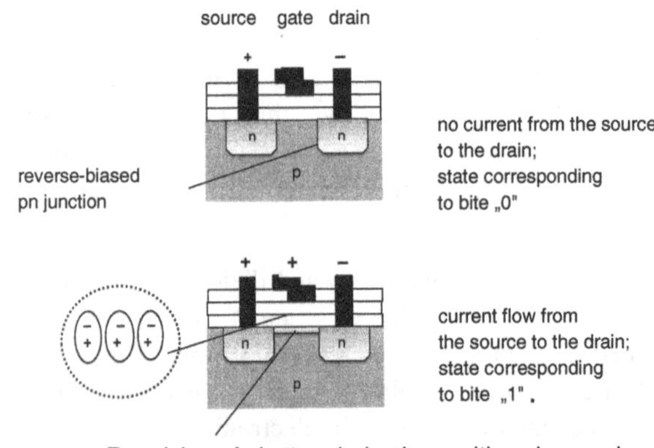

Fig. 1.8 Diagram of a metal-oxide-semiconductor field effect transistor (MOSDFET). It includes a capacitor-like gate composed of a metallic electrode, a dielectric and a semiconductor layer. Voltage switching at the gate generates either no current flow or current flow between the source and drain terminals; this corresponds to a sequence of '0' and '1' bits in the binary system. Transistors with MOSFET structure are the most commonly used active components of DRAMs (dynamic random access memories) in computers and intelligent cards

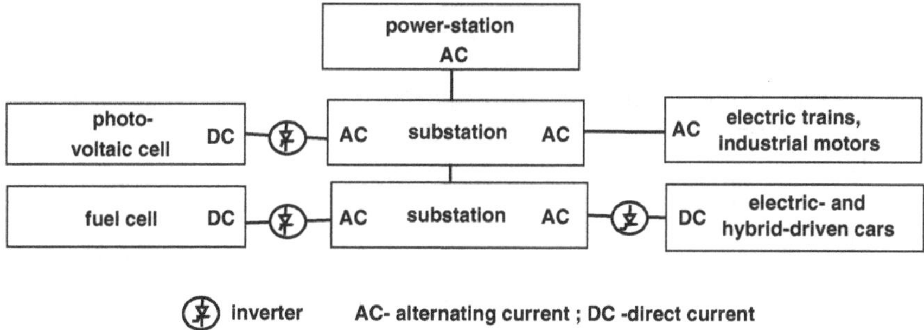

Fig. 1.9 Diagram of modern grid of power transmission lines

Ceramics in Power Grids Active semiconductor devices have become also an integral part of power grids. Many devices receiving and storing electric energy and sources supporting its production (photovoltaic and fuel cells) generate direct current (DC), whereas in the power grid high-voltage alternating current (AC), produced in power plants, is transmitted. In addition to rectifiers in the grids, inverters—devices converting DC to AC with adjustable frequency—are therefore needed (Fig. 1.9). In these applications very high breakthrough resistance is required. Such properties are exhibited by certain polymorphic forms of silicon carbide (e.g. 4H SiC).

Light Emitting Diodes Another dynamically growing application for semiconductors (these with direct energy band gap) such as GaAs, GaN is lighting. In light-emitting diodes (LEDs) made of such semiconductors, the injection of electrons from an n-type semiconductor to a p-type, induced by a near external electromagnetic field, sustains their recombination with electron holes, giving rise to continuous emission of photons in the visible light spectrum (→ 6. Materials versus light). Electric energy is converted here to light with nearly 100 % efficiency, LEDs are of miniature sizes, of the order of millimetres, since, due to the opacity of semiconductors, light photons are emitted only from near-surface layers. Thanks to these properties, LEDs are increasingly used for lighting purposes, from back-lighting of miniature mobile-phone displays to illuminating streets and buildings.

The trend to miniaturise semiconductor devices continues. It is believed that, a broader use of ferroelectrics, carbon nanotubes and grapheme (→ 5. Unusual dielectrics and ceramic conductors), further miniaturisation and ease of switching of semiconductor transistors and diodes can be made possible.

Glass Optical Fibers While semiconductors are opaque to visible light, inherently transparent materials predominate in ceramic dielectrics. The flagship material in this category is oxide (silica) glass which has for years been used for the production of prisms, lenses and optical windows. In the post-industrial epoch, oxide glass is beginning to be used for making optical fibres acting as light pipes: 'the light goes where the fibre goes' (→ 2. Tradition continued. Silicate tradition; 6. Materials versus light). The total diameter of their core and sheath typically does not exceed

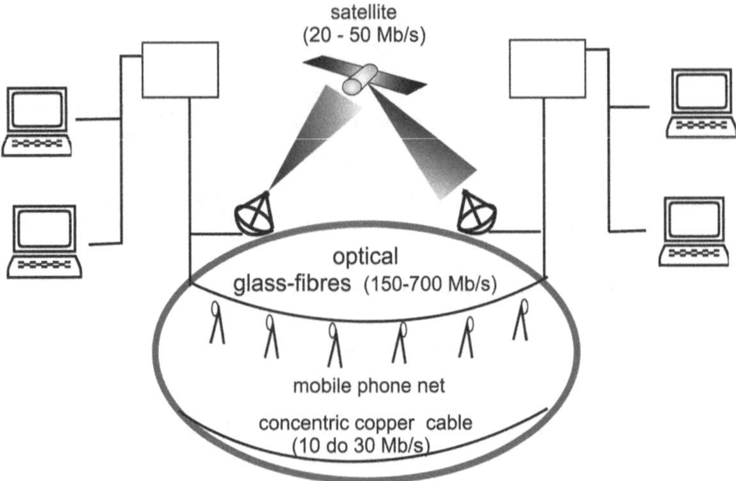

Fig. 1.10 Digital data flow in various information technology (IT) and cellular networks. For each type of network, the bandwidth is given (i.e. the maximum number of bits per second that can be transferred through the network)

125 μm, making it possible to produce cables with a large number of fibres. Moreover, there are no interactions between light waves of different frequency and amplitudes. In a single fibre, from a few to several dozen light waves can therefore be multiplexed (braided together). All this enables information technology (IT) networks made of glass fibres to offer the highest bandwidth, i.e. the number of bits per second (Fig. 1.10).

Pyroelectric, Piezoelectric and Ferroelectric Ceramics The category of transparent materials includes dielectrics where an application of heat, stress and external electromagnetic field brings about substantial pyroelectric, piezoelectric and ferroelectric effects (→ 5. Unusual dielectrics and conductors). They are used in many applications, including sensors and acruators of active intelligent systems (→ 7. Imitating and supporting nature), dynamic random access memories (DRAM), smart cards and fast telecom equipment. An important role is played by oxides with perovskite structures, such as $BaTiO_3$ and PZT, i.e. lead titanate zirconate [$Pb(Zr_yTi_{1-y})O_3$] and PZLT, derived from PZT on Pb-cation- substitution by lanthanum (La). At present, complex oxides free from harmful lead, such as bismuth ferrite ($BiFeO_3$) or niobates [$LiNbO_3$, $(Na,K)NbO_3$] are arousing interest.

Solid Electrolytes Versus Fuel Cells Ceramic solid electrolytes are used in the production of electricity in high-temperature solid oxide fuel cells (SOFC), in which electrochemical combustion reactions occur. With around 60 % efficiency of generation (compared to 35 % in conventional methods), the SOFC have already become a useful secondary source of electric energy. Single cells are, as a rule, miniature devices (Fig. 1.11). In local power plants with power outputs of 1–10 MW

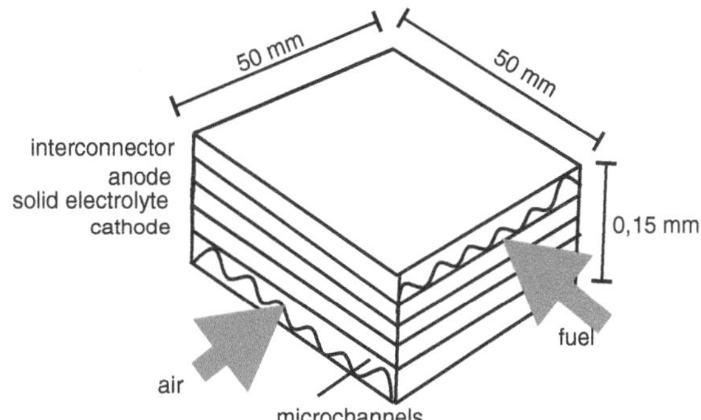

Fig. 1.11 Structural diagram of a single solid oxide fuel cell (operating at an elevated —temperature)

which are already being built, current generators are therefore assembled via the modular composition of a large number of stacks of such cells.

High-Temperature Superconductors A practical use of no-resistance flow of electric current through high-temperature ceramic superconductors (→ 5. Unusual dielectrics and ceramic conductors) is now starting. Although the mechanism of superconductivity has not yet been investigated, it is known that in the structure of the most typical superconductors, namely cuprates (→ Fig. 5.10) the movement of electrons, which contribute to superconductivity, is parallel to the Cu–O planes of their structure. Therefore, for highly efficient use of superconductivity, favourable orientation of those planes in parallel to the current transport direction in a cable, is required. This is achieved by applying epitaxial layers of superconductors on a substrate (MgO; Ni) by vapour deposition from the gas phase.

Chapter 2
The Tradition Continued

The materials mentioned here complete the chain of ceramic materials produced since the dawn of humanity. They are mass-produced which is enabled by use of, little processed natural raw materials, having appropriate immanent properties. One can distinguish materials based on the traditions of using clay, lime and silica as well as of refractory rocks (→Fig. 1.1).

2.1 The Clay Tradition

Clay-Water Mix This tradition includes products for which the basic raw material is a sedimentary rock: clay. The main components of clay are clay minerals having the structure of phyllosilicates. A typical phyllosilicate is kaolinite, $Al_4[Si_4O_{10}](OH)_8$. Its atomic structure is composed of strongly bound silicon oxide and gibbsite layers (see insert in Fig. 2.1). As a result of their origin, clays contain various amounts of quartz (SiO_2) and feldspar, the most common minerals in the crust (i.e. $KAlSi_3O_8$, $NaAlSi_3O_8$, $CaAl_2Si_2O_8$); and fragments of other rocks and organic substances. Kaolinite-rich clays of a low impurity contents are called kaolin.

The specific properties of clay minerals-water masses have been exploited from the beginning. Clay minerals form agglomerates composed of 2–5 crystallites with thicknesses of 2–8 nm and widths <100 nm. As a result of Coulombic attraction of hydroxide layers in crystallites and water molecules, the agglomerates are wetted by water, which forms thin layers on their surface. This results in the formation of water bridges between crystallites (Fig. 2.1). Since, the wetting angle (→Fig. 4.2 in Chap. 4) is less than 90°, the meniscus (i.e. the water-air interface) of the water bridges is concave and, due to the low thickness of the water layers, has a very small radius of curvature.

There is a relationship between the chemical potential of atoms and surface curvature:

$$\Delta\mu = \gamma\Omega\left(\frac{1}{a_1} + \frac{1}{a_2}\right) \tag{2.1}$$

R. Pampuch, *An Introduction to Ceramics*,
Lecture Notes in Chemistry 86, DOI 10.1007/978-3-319-10410-2_2

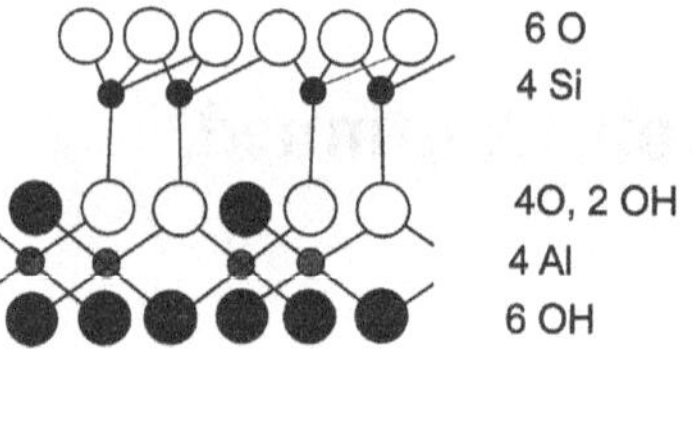

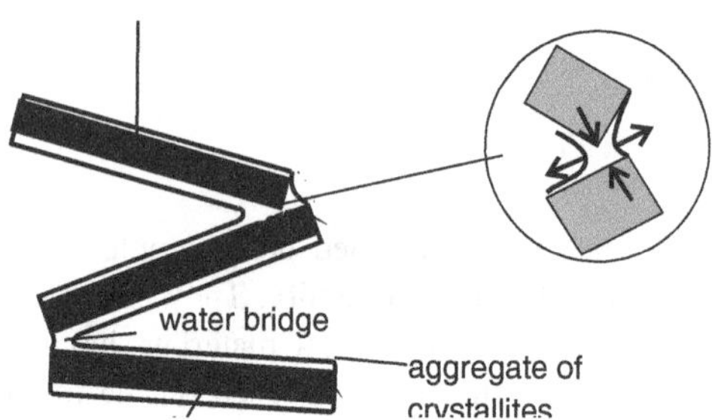

Fig. 2.1 Agglomerate of clay minerals crystallites with adsorbed water layers. *Top* insert: cell of kaolinite structure; *right* insert: capillary forces in water bridges between the crystallites

where: $\Delta\mu$—the chemical potential difference of atoms on curved and flat surfaces; γ—surface tension at the interface (here at the water-air interface); Ω—atomic volume; a_1 and a_2—two basic curvature radii, whereby $a_i > 0$ for a convex surface and $a_i < 0$ for a concave surface. According to Eq. (2.1), atoms on the convex side have higher and on the concave side lower chemical potential than on a flat surface. A chemical potential gradient is therefore formed across the interface. Since grad $(\mu) = -\boldsymbol{F}$, where $\boldsymbol{F}$ is force, forces, called capillary, arise at the interface. For water surface tension $\gamma = 7.2 \cdot 10^{-2}$ [N·m^{-1}] and a_1 equal to several nm, the capillary force can be as high as $1 \cdot 10^3$ Pa.

In equilibrium, the capillary forces in the water bridges are balanced by identical compressive forces acting perpendicularly to them (Fig. 2.2). This causes agglomerates of clay mineral crystallites to be strongly attracted to one another.

To lower the resulting cohesion of crystallites more water must be added and to initiate deformation of the clay in this state stress has to be applied. On increasing the applied stress above a limit value (plasticity limit) the 'lubricating' action of water facilitates a further increase in deformation. After the load is relieved, the capillary forces cause the clay-water mix to retain the deformation from the load. This is referred to as plastic behaviour.

Deformation of material by shearing requires, as a rule, the application of a lower stress than for other modes of applying load. Such a deformation can be achieved at many points of the clay-water mix, if tangent stress is applied to its surface. The discovery of this behaviour of the clay-water mix is believed to have been the reason behind an innovation introduced around 5,000 years ago in Sumer, namely, the formation of clay products with a potter's wheel.

Slurries Agglomerates of clay minerals' crystallites can be, under some conditions, permanently dispersed in water, forming suspensions, also referred to as slips or slurries. There is an evidence for the stabilisation of slurries through the use of

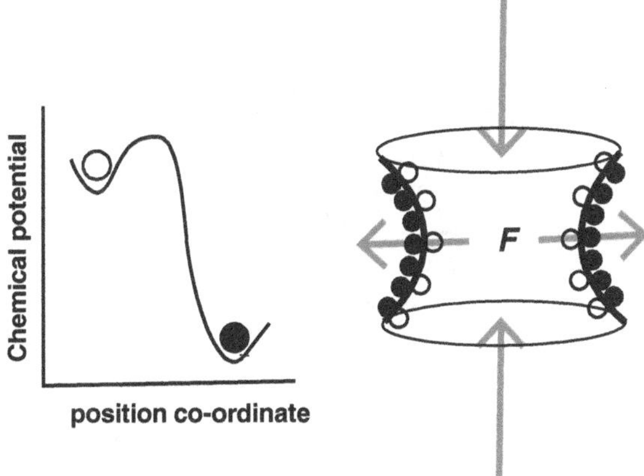

Fig. 2.2 Chemical potential of atoms at the convex and concave sides of a liquid-gas interface (*left-hand schema*) and compressive forces due to capillary forces in the water bridges (*right-hand schema*)

alkali-rich plant ash as early as the Hellenistic era, i.e. around 2,300 years ago. We know now that this is due to positive influence of K^+ or Na^+ present in the water solution. The durability of slurry is, namely, explained by sorption of these alkaline cations onto negatively charged sites in clay minerals' crystallites. Achieving in this way identical positive surface charges, the crystallite agglomerates repel one another, preventing their flocculation. However, the effect of cations being surrounded with a hydration layer competes with the tendency of their direct sorption. Today it is known that the effectiveness of K^+ or Na^+ results from their tendency to become surrounded with thin hydration layers only. With insignificant concentrations, e.g. between 10^{-2} and 10^{-4} M, of potassium or sodium cations in the water solution, a double electric layer is formed around the crystallite aggregates (Fig. 2.3a). A certain number of cations of this layer are strongly bound to the aggregate surface; the remaining part of that layer, however, can easily move under the shear load. Stable slurries can also be created by adsorbing on the surface of the aggregates organic molecules which form steric hindrances against an approach of the aggregates (Fig. 2.3b).

Phase Transformations During Firing The finds from the peak of development of the Neolithic revolution bear witness to, discovery at this time that the clay-water mix, when fired at temperatures achievable when burning wood, turns into a solid and strong body. We know now more about the transformations which occur in kaolinite-feldspar—quartz mixes during firing (Fig. 2.4). This knowledge permits to conclude that a decisive role it formation of the strong body plays a highly viscous SiO_2 and Al_2O_3-rich liquid appearing about 800 °C. Wetting solid particles, the liquid forms bridges between them (Fig. 2.5) is and on this way capillary forces are aroused (→Fig. 2.2) owing to which the system is in a state corresponding to application of external hydrostatic compressive stress. The stress is considered to be the driving force of sintering, i.e. increase of packing density and cohesion of the particles at elevated temperature. The action of the SiO_2- and Al_2O_3-rich liquid is very effective in sintering, as testified by a high strength of the porous body of

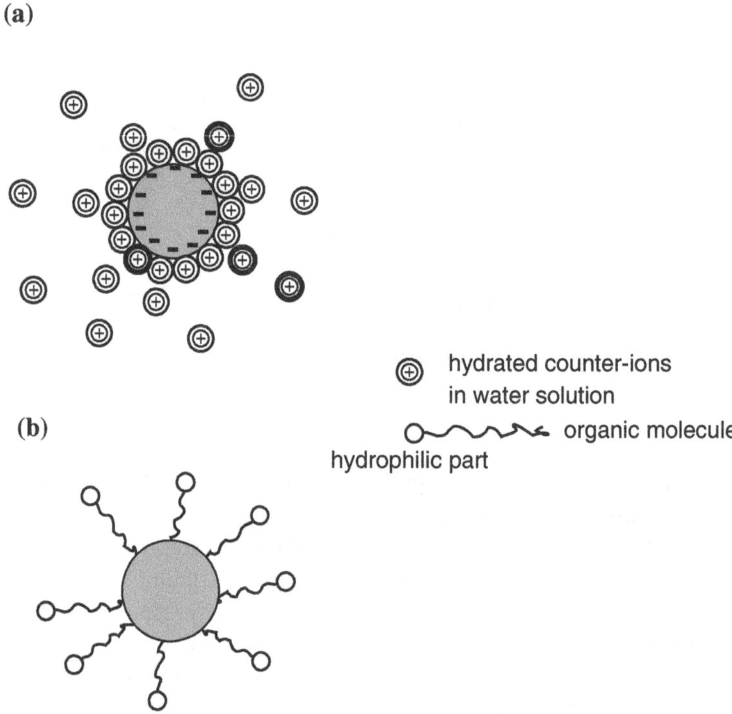

Fig. 2.3 Stabilisation of suspensions of solid particles in water: creation of a double layer by sorption of hydrated anti-ions (**a**); creation of steric hindrances by sorption of organic chain molecules with hydrophobic and hydrophilic functional groups (**b**)

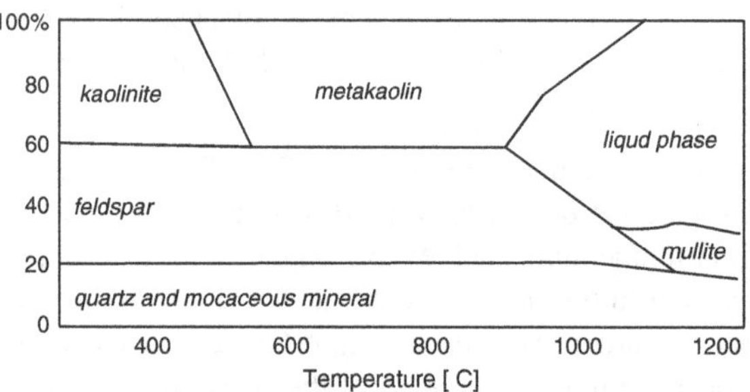

Fig. 2.4 Phase transformations during firing of: kaolinite [$Al_4[Si_4O_{10}](OH_9))$]—feldspars [K,Na) $AlSi_3O_8$-$CaAl_2Si_3O_8$-] quartz [SiO_2] mixeses (generalised)

earthenware. A less porous microstructure is obtained by using a higher percentage of K and Na-containing feldspars in the mix and higher temperature of firing. Both factors increase the amount of the liquid phase formed on firing and its viscosity. Owing to this the liquid nearly completely fills up the pores. In this way, the

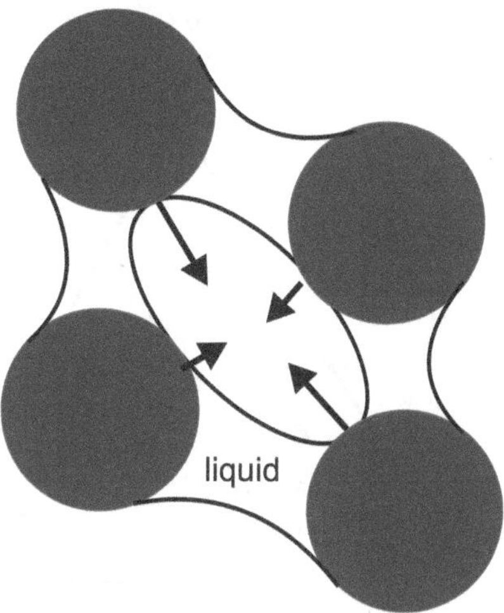

Fig. 2.5 Porous system of solid particles wetted by a liquid at an elevated temperature. Due to capillary forces arising in liquid bridges between the particles, the system is in a state corresponding application of an external hydrostatic compressive stress which is the driving force. For sintering increasing the packing density of particles with elimination of pores

microstructure of porcelain body is formed, which is dense and impermeable to water and gases, and transparent in thin-walled products. Because relatively pure raw materials, like kaolin, are used—the porcelain body is more or less white. Most of the liquid phase solidifies on cooling as brittle glassy phase (→next section) but this is compensated by an enhanced crystallisation of secondary mullite (composition close to $Al_6Si_2O_{13}$). Secondary mullite crystallises in form of needles of a high strength, and their grid reinforces the brittle glassy phase.

The composition of raw materials mixes used to obtain the mentioned products and stoneware, composed of still more glassy phase than porcelain, is shown in Fig. 2.6.

2.2 The Silica Tradition

Oxide Glass The formation of melts solidifying into a vitreous, amorphous phase on cooling down has been discovered already in ancient times. Mixtures of solid components, containing, as we know today, around 70 % SiO_2 and 30 % $Na_2O + CaO$, were molten at temperatures achievable when burning wood to obtain upon cooling a transparent material: glass. The chemical composition of today's standard oxide glass, such as [silica] soda-lime glass, does not differ significantly from that of antique glass (Table 2.1). However, for special applications, other

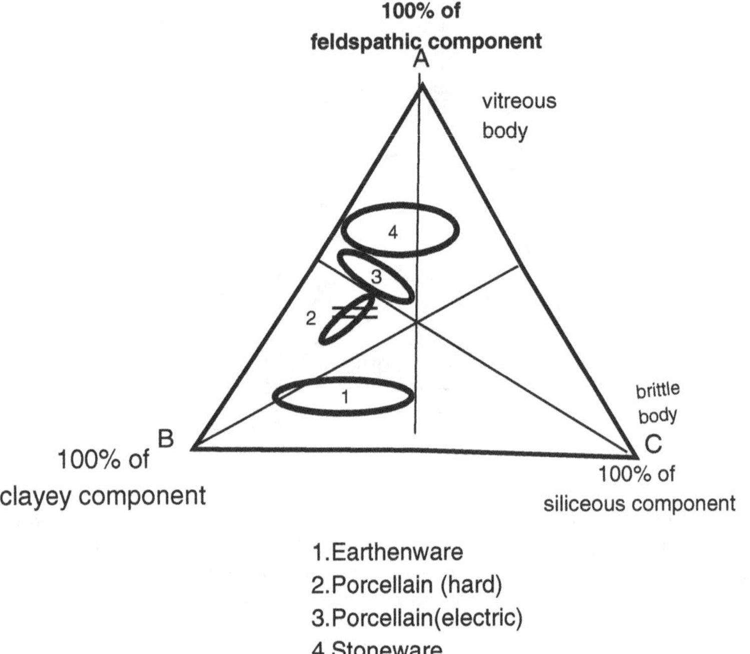

Fig. 2.6 Typical compositions used for production of earthenware, porcelain and stoneware

Table 2.1 Composition of antique and modern window and pottery glass types

	SiO_2	Al_2O_3	CaO	MgO	K_2O	Na_2O	B_2O_3
Babylonian glass (fifteenth to fourteenth century BC)	64	1	6	6	5	18	
Egyptian glass (fourteenth to eleventh century BC)	66	1	6	5	3	19	
Modern window and pottery glass	73–74	0–1	9–10	1	1	15	<1

types of oxide glass have developed over time. These include borosilicate glass, in which the addition of a small amount of alkalis to SiO_2 and B_2O_3 makes it possible to obtain products with a thermal expansion coefficient ($\alpha = 3–6 \cdot 10^{-6}$ °C^{-1}) lower than that of soda-lime glass ($\alpha = 1 \cdot 10^{-5}$ °C^{-1}) and with greater resistance to an aggressive environment. Furthermore, lead glass which contains SiO_2 and PbO, along with small shares of sodium and potassium, is characterised by a high gloss and can be formed into various shapes within a wider temperature range. These properties are used, for example, to manufacture decorative glass products ('crystal' glass), as well as in electronics. The oxide glass family also includes aluminosilicate glass (called E or S glass), which includes an increased proportion of Al_2O_3, MgO and CaO. Such glass is characterised by mechanical properties superior to those of most other glass types (Fig. 2.7).

Non-oxide Glass The tendency to form amorphous solids like the oxide glasses is not limited to oxygenous systems. It also occurs when oxygen is replaced with fluorine and, at the same time, Si, Ge or B are replaced or supplemented by other elements. In this way are obtained, for example, fluoride glasses, containing 57 % molar ZrF_4, 36 % BF_2, 4 % LaF_3 and 3 % AlF_3). Chalcogenide glass, containing e.g. molar percentages of 48 % Te, 30 % As, 12 % Si and 10 % Ge, have specific properties which distinguish them from oxide glass. Namely, a high-crystallisation rate which may be initiated by a laser pulse within only 30 or so nanoseconds and very good transparency to infrared radiation (wavelength range from 0.6 to 15 μm. The electrical conductivity and reflectance (→5. Materials versus light) of the crystalline and amorphous chalcogenide differ significantly. Through appropriate heating and cooling cycles, point sequences can be created with stronger and weaker light reflection, corresponding, respectively, to bits = 1 and = 0 in the binary mathematical system. This makes it possible to save and read binary information on CD-ROMs and DVDs.

Glass Structure Glass formation has been better, if only phenomenologically, understood now than in ancient times. In the case of liquid metal alloys and ionic substances, a first-order transformation usually occurs while cooling at a relatively low rate. At a specific temperature it causes transition of the liquid into the thermodynamically more stable crystalline phase. In oxide liquids which can be transformed into glass (typical are liquids containing around 70 % SiO_2 and 30 % Na_2O + CaO), the first-order transformation does not occur. However, at a temperature referred to as the glass point, a second-order transformation is observed, during which the liquid gradually passes to a solid with an amorphous structure, presumably inherited to a considerable degree from the liquid Although in the absence of direct methods it is difficult to speak of details, the lack of x-ray diffraction in glass proves that, in contrast to crystalline structures, there is no

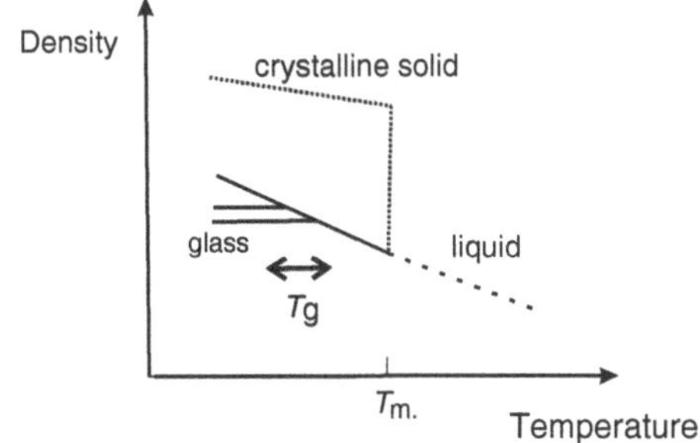

Fig. 2.7 Density changes at on a first-order liquid-crystalline solid phase transition and on attaining the glass point by a liquid

long-range order in glass, but at most short-range order. The possible atomic structure of crystalline and amorphous silicon dioxide is illustrated in Fig. 2.8, assuming that it is formed in both cases by tetrahedral [SiO_4] groups, bound together by common oxygen atoms, called oxygen bridges. While there is short-range order in the tetrahedral groups, their network in amorphous form of SiO_2 lacks long-range order (Fig. 2.8b).

The amorphous network can be modified by Al^{3+} and Ti^{4+}, and by Ca^{2+}, Li^+, Na^+, K^+. Three probable types of reactions of the cations, introduced in form of oxides, with the silicon-oxygen network are: oxygen donation to the silicon-oxygen network causing breaking of the oxygen bridges (Fig. 2.9a); oxygen donation increasing the number of oxygen atoms coordinated with the central silicon atom (Fig. 2.9b); a combination of both reactions (Fig. 2.9c). The type of reaction may be assumed to depend on the cation-oxygen Coulombic attraction, which is proportional to the ratio of the cationic charge to its radius, z/r. For Si^{4+} coordinating 4 anions of oxygen, $z/r = 1.6 \cdot 10^{-8}$ [$C \cdot m^{-1}$], which contrasts with $z/r = 0.1 \cdot 10^{-8}$ [$C \cdot m^{-1}$] in the case of K^+ and Na^+ and $z/r = 0.3 \cdot 10^{-8}$ [$C \cdot m^{-1}$] at Ca^{2+}, Mg^{2+} and Li^+. $^+$. Therefore, it can be assumed that when introducing potassium oxides or sodium to the silicon-oxygen melt, the reaction illustrated in Fig. 2.9a dominates.

TTT Graphs These and other considerations, however, are more or less speculative so it is worth to keep on the phenomenological aspects of glass formation. To this end, a TTT (time-temperature-transformation) graph will be used (Fig. 2.10). The shadowed area in Fig. 2.10 corresponds to time-temperature conditions in which more than a 1 % of the liquid is crystallising when a melt is cooled from a higher temperature. Therefore, a liquid at initial temperature T_p will not be crystallising as long as it is cooled at a higher rate than minimum, given by:

$$dT/dt_{min} \approx \left(T_p - T_{min}\right)/t_{min} \tag{2.2}$$

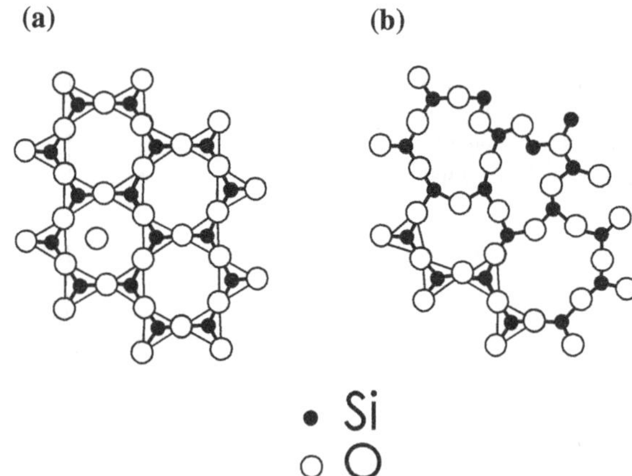

Fig. 2.8 Two-dimensional schema of a crystalline structure **a** and amorphous network in glass **b** composed of [SiO_4] tetrahedral coordination groups

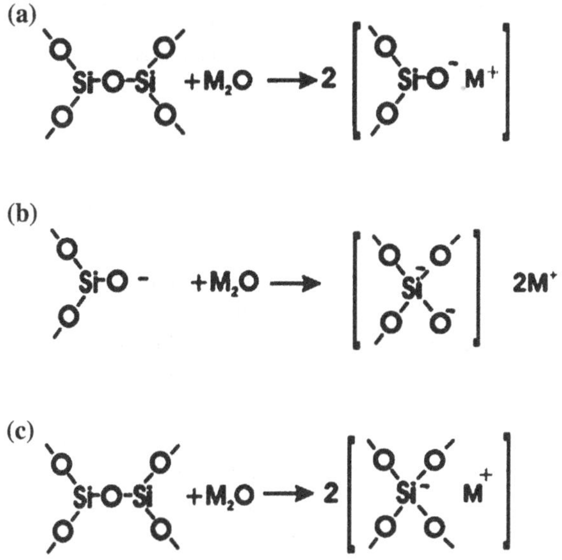

Fig. 2.9 Probable reactions of oxides with amorphous silicon-oxygen network. **a** Donation of oxygen to the silicate network resulting in breaking down of oxygenbridges in the network. **b** Donation of oxygen to the silicate network resulting in an increasednumber of oxygen atoms coordinated with silicon. **c** Combination of reactions **a** and **b**

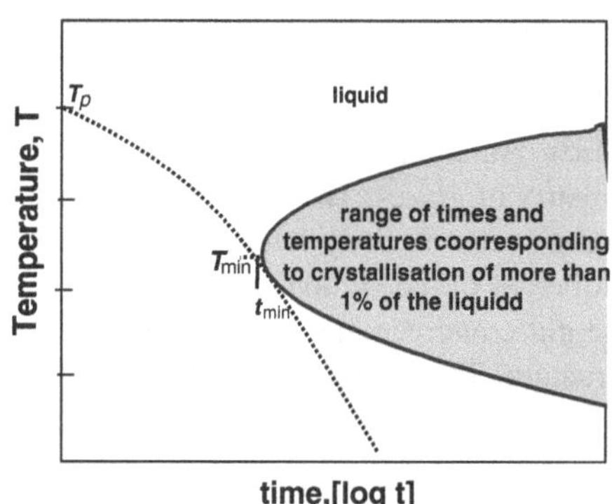

Fig. 2.10 Temperature-time-transformation (TTT) graph, typical for a soda-lime glass composition. The application the TTT graph is explained in the text

Minimum Cooling Rates for Glass Formation where: T_p initial temperature of the liquid; the temperature; the values of T_{min} and time t_{min}—are Indicated on the graph. The minimum cooling rate corresponds to the dotted line in Fig. 2.10. Table 2.2 specifies the minimum cooling rates determined experimentally for various systems. These constitute a measure of the tendency of various liquids to making glass while being cooled. In contrast to many silicon-oxygenous system, to produce metallic glass, extremely high cooling rate is necessary.

Devitrificates Glass products with specific properties include devitrificates. They are typically composed of a 30–90 % volume of small crystalline grains of ceramic phases bonded with non-transformed glass. Generally speaking, devitrificates have the characteristics of both crystalline and amorphous phases, which can be changed over a broad range. Compared to glass, the presence of crystalline phases in

Table 2.2 Approximate minimum cooling rates for liquids required to create an amorphous vitreous phase

Substance	dT/dt [°C·s^{-1}]
Soda-lime glass	$1 \cdot 10^{-5}$
SiO_2	$1 \cdot 10^{-4}$
As_2S_3	$1 \cdot 10^{-1}$
metals	1.108

devitrificates reduces, for example, the value of the thermal expansion coefficient and thus increases resistance to thermal shocks (see further text on refractory tradition) while also increasing mechanical strength and resistance to brittle fracture.

The formation of ceramic crystalline phases in devitrificates, ceramisation, is performed through controlled crystallisation of a heated glassy mass. The continuous network typical for glass has to be modified to allow the formation and growth of the crystalline phase. In case of silicate glasses one method is to introduce suitable network-modifier cations to the liquid which, by initiating the reactions illustrated in Fig. 2.9, causes the Si-O-Si bonds to break and smaller silicon-oxygen domains to form. This, in turn, causes increased mobility of domains, which promotes the crystallisation. An important role is also played by the presence of interfaces. At interfaces, the potential energy barrier for crystallisation is lower than in the bulk and (heterogeneous) nucleation is facilitated. These requirements are met only by some systems, for example, those shown in Table 2.3.

Glaze An amorphous silicon-oxygen vitreous phase is the main constituents of the t matrix of glazes. The term glaze is used to describe vitreous layers applied to the porous ceramic body to strengthen, waterproof and decorate it. Accordingly, their properties of the vitreous layers have in first place to be adjusted to the body, so that the glaze and ceramic body form a coherent, mechanically and thermally resistant structure. This applies in particular to adjusting the thermal expansion coefficient,

Table 2.3 Some three- and four-component systems used for the fabrication of devitrificates and the crystalline phase in devitrificates

System	Crystalline phase	
$CaO–Al_2O_3–SiO_2$	Anorthite	$CaAl_2Si_2O_8$
$MgO–Al_2O_3–SiO_2$	Cordierite	$Mg_3Al_4Si_5O_{18}$
	Spinel	$MgAl_2O_4$
$MgO–CaO–Al_2O_3–SiO_2$	Anorthite	$CaAl_2Si_2O_8$
	Cordierite	$Mg_3Al_4Si_5O_{18}$
	Spinel	$MgAl_2O_4$
	Diopside	$CaMgSi_2O_6$
$ZnO–Al_2O_3–SiO_2$	Gahnite	$ZnAl_2O_4$
	Willemite	Zn_2SiO_4
$Li_2O–Al_2O_3–SiO_2$	Spodumene	$LiAlSi_2O_6$
	Eucryptite	$LiAlSiO_4$

which enables reduction of residual thermal stress (→3. Ceramics to overcome brittleness) at the body-glaze interface. Recent developments include a transformation of the amorphous glaze into polycrystalline devitrificates (see previous text) which makes it more mechanically strong and resistant to thermal shocks, and making TiO_2-rich glazes, to activate photocatalysis in which light photons generate bactericidal radicals (→1. A brief history of ceramic innovation). Other issues, related to the optical properties of glazes, shall be discussed in more detail in an appropriate context (→6. Materials versus light).

2.3 The Lime Tradition

Properties of Ca–O Groups The term used here refers to the fact that calcium compounds, such as the gypsum ($CaSO_4{\cdot}2H_2O$), quicklime (CaO), or calcium silicates that form the bulk of Portland cement clinker, have played a key role in the field of construction materials. The latter include mortars for bonding building construction blocks and the blocks themselves, followed by decorative moulding, drywall panels for partition walls, ceilings a.s.o. The use of, gypsum and sand mortars was reported as early as in ancient Egyptian times, and quicklime mortars were used on a wider scale not only in the Roman Empire, but also in earlier structures in Minoan Crete.

These applications result from specific properties of Ca–O groups, namely, their high reactivity with CO_2 and water. Reactivity with CO_2 is used to obtain non-hydraulic lime mortars in a cycle of reactions shown in Fig. 2.11. Natural limestone is converted to reactive quicklime(CaO) which, when mixed with water, forms, after some time, a thick slurry that can be applied on construction blocks to bond

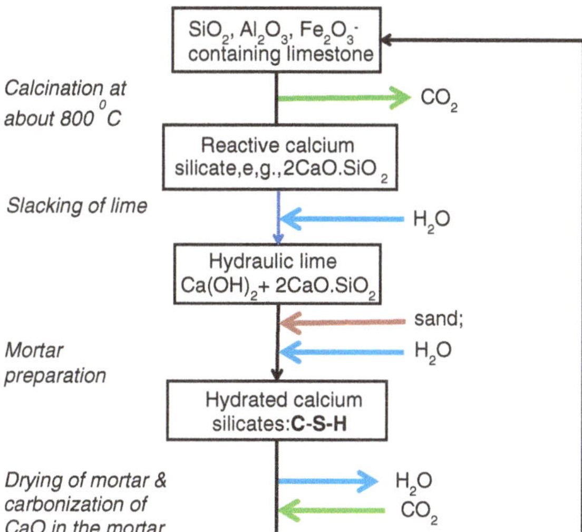

Fig. 2.11 Cycle of reactions used in the case of non-hydraulic lime mortars

them together; the slurry, reacting with CO_2 from the atmosphere, turns back into a layer of limestone of moderate strength.

Portland Cement The most widespread material exploiting the unique properties of calcium compounds has been, for over 150 years, hydraulic Portland cement, which sets after mixing with water and binds a mineral aggregate. Therefore, it can be used to make concrete, which became an indispensable construction material for the foundations of buildings and dams, and an important material in road, bridge and building construction. A wide variety of concrete types have been created for different applications. Let us mention three of them. Standard concrete is a mixture of hydraulic cement, fine and/or coarse mineral aggregate and possibly some other additives (up to 5 %). At some point after mixing with water and compacting with vibrating equipment, durable concrete is formed. At the end of the twentieth century, self-compacting concrete was developed—a mixture of cement, aggregate and various chemicals with high initial liquidity which, when cast in any shape, compacts itself under its own weight. Another product is cellular concrete, made by subjecting the mix to steam which causes the simultaneous formation of a large number of pores and setting of cement.

The intermediate product in the production of concrete is so-called Portland cement clinker, a clinker made by firing marl (clay, containing calcium carbonate and quartz sand) at a temperature of about 1,450 °C. The typical phase composition of the clinker is shown in Table 2.4. The cement used to form a water suspension (cement paste) also contains gypsum to slow down the hydration and setting processes, thus preventing significant internal stress and cracking of the solidified hydration product. Moreover, high-strength cement contains typically small additions of silica dust to reduce the crystallisation of calcium hydroxide in the hydration product.

The hydration mechanism was phenomenologically investigated mainly for alite ($3CaO \cdot SiO_2$, C_3S), which forms 55–75 % of Portland cement clinker. This process depends on the growth of a water-insoluble hydration product, which is calcium silicate hydrate, or C-S-H (the so-called tobermorite-like or jennite-like phase). The growth involves successive addition to two-dimensional silicon-oxygen chains, with a cross-section diameter of several nanometres, of $[SiO_4]$ coordination tetrahedra, Ca^{2+} ions and H_2O molecules (Fig. 2.12). The reasons for the growth of such fibrous phases are not clear and the explanations are controversial. More understandable are

Table 2.4 Typical phase composition of Portland cement clinker[a]

Phase	Percentage
$3CaO \cdot 2SiO_2$ (C_3S, alite)	55–75
$2CaO \cdot SiO_2$ (C_2S, belite)	<25
$3CaO \cdot Al_2O_3$ (C_3A)	<12
$4CaO \cdot Al_2O_3 \cdot Fe_2O_2$ (C_3AF)	<8

[a] In addition to ground clinker, the Portland cement contains ca 3.5 % $CaSO_4 \cdot H_2O$ to slow down the reaction of C_3A with water

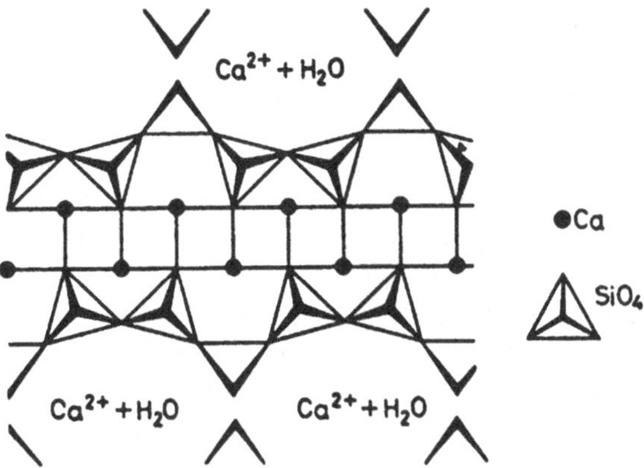

Fig. 2.12 Structure of tobermorite- and jennite-phases formed during hydration of alite (schematic)

the results: he hardening of the alite-water systems due to growth of the two-dimensional tobermorite- and jennite-like phase in the water space. The fibrous matter branch off, forming a tangled net, the branching probably due to the internal stress caused by defects. Given the considerable surface area, the cohesion of these fibrous matters should be substantial, which would reasonably explain the great strength (particularly compressive) of the hydration product. This product also contains numerous pores with radii of ca 2 nm and less frequent voids with larger diameters. In classic cement, about 10 % of its volume consists of $Ca(OH)_2$ particles larger than 1 mm as well as particles of aluminosilicate phases. To some degree, these particles compensate for the effect of the fibrous net formation, hence the addition of silica dust in high-strength Portland cement products.

Refractory Cements There are also other types of hydraulic cements. An example is a cement produced since the mid-twentieth century whose main ingredients are the calcium aluminates $CaO{\cdot}Al_2O_3$ (CA) and 2 $CaO{\cdot}7Al_2O_3$. These compounds react exothermically with water. An increase in temperature due to the reaction causes the rapid dissolution of anhydrous aluminates in water and the subsequent precipitation of hydrated aluminates from the supersaturated solution in form of a tangled net. This results, at first, in the formation of a thick paste, and then of a solid with significant strength. In addition to a high setting (hardening) rate, such cements are characterised by high resistance to high temperatures. Therefore, they are increasingly used as refractory compounds for the lining of high-temperature equipment, and are often called refractory cements.

In the twenty-first century, attention has been drawn to materials, hardening within several minutes at a temperature about 100 °C. However, this is but based on a different mechanism than in the case of lime, Portland cement or refractory cements. These are so-called geopolymers, which constitute a category of inorganic polymers, with the nominal composition $ySiO_2{\cdot}Al_2O_3{\cdot}M_2O$ (M = Na, K or Cs, $3 < y < 35$. Their structure is formed most probably of nanocrystallites with a zeolite-like structure (→Fig. 4.6), composed of $[SiO_4]$ and $[AlO_4]$ coordination polyhedra, bonded together via oxygen bridges. The positive charge of Al^{3+}, smaller than that of Si, is compensated by the presence of alkaline cations, such as K, Na or Li.

The precursors of the geopolymers are mostly waste pozzolanic materials, which include metakaolin (→Fig. 2.4), volatile ash and blast furnace slag. They ensure the rapid and easy dissolution of SiO_2 and Al_2O_3 in highly alkaline water solutions, e.g. in hydroxide Na(OH) or sodium metasilicate Na_2SiO_3—so-called water glass. As a result sols are created in which monomer hydroxides of the Si-O-Al-O-[poly(silalane)], -Si-O-Al-O-Si-O-[poly(silalane-siloxide)] type are found. Those monomers are polycondensed at moderate temperatures (up to 100 °C) according to the reaction: R-OH- + -HO- R = R-O-R + H_2O.

The driving force of geopolymer applications is environmental protection. First, waste material is used to produce them. Secondly, it is important that considerably less CO_2 is released into the atmosphere during their production than in the case of Portland cement. The properties of solidified products of geopolymers are also important. While the properties of Portland cement deteriorate rapidly if the temperature is increased to about 400 °C, the mechanical strength of the products of geopolymer setting is retained up to 800–1,200 °C, owing to their transformation into thermally resistant aluminosilicates such as nepheline $KNa_3[AlSiO_4]_4$, calsilite $K[AlSiO_4]$, leucite ($K[AlSi_2O_6]$) and pollucite $Cs[AlSi_2O_6]$.

2.4 The Refractory Tradition

Microstructure and Properties Producing and handling molten and hot metals, melting of glass-forming compositions, Portland cement clinker and coke production waste incineration a require constructions (furnaces, kilns, ovens) clad with refractories, i.e. materials that can withstand temperatures from 1,200 to 1,500 °C . Although bloomery-type furnaces (pits dug in clay) had been used for centuries, the mass production of refractories started only during the Industrial Revolution. Because of great market demand (→Table 1.1), both early and contemporary refractory materials have been made mainly of low-modified natural raw materials. Due to the high operating temperatures of refractory materials, these include substances whose mineral ingredients are melting above 1,700 °C (SiO_2), 2,000 °C (αAl_2O_3) or 2,500 °C (Cr_2O_3, MgO, CaO) They are used in the form of pastes or components made by pressure-forming and sintering at a high temperature. With raw materials comminuted to particles of <1 mm, a total porosity of 40 % can be achieved and further reduced by compressing. On heating the pulverised raw materials, liquid eutectics are formed due to the presence of silica-rich impurities, and liquid bridges created between the solid grains. In a way shown in Fig. 2.5 the contact area and cohesion of grains are increase to a higher level, sufficient for refractory materials performance under their own weight. The chemical composition and the porous microstructure permit to achieve:

1. retention of their form an integrity under their own weight at temperatures up to 1,500 °C (or even higher);
2. resistance to high-temperature corrosion in aggressive environment (liquids, gases and slag);
3. low thermal conductivity; which ensures thermal insulation of the hot interior of high-temperature furnaces and reactors;
4. resistance to sudden temperature changes (thermal shocks), to which many refractory linings are subjected.

High-Temperature Corrosion Let us make more detailed comments. High-temperature corrosion is mainly due to reactions between refractory materials and the aggressive environment. In addition to reactions with aggressive gases, these are mainly reactions with molten metals, glass and particularly with slag floating on the surface of melted metal or glass. Such reactions may cause transformations in the material with loss of weight and deterioration of properties.

In the case of refractory materials made of oxide ingredients, the reactions are mainly an acid-base ones; in materials based on SiC and Si_3N_4, oxidation-reduction reactions also take place. In general, the intensity and rate of the acid-base reactions is highest when one of the substances involved is acidic and the other basic. The rate decreases when the acidity or basicity of the substances is similar. This is illustrated in the example in Table 2.5, in which shown is the dissolution rate of an acidic refractory material in a molten basic FeO.

Magnesite (based on MgO), chromium-magnesite (based on Cr_2O_3 and MgO), dolomite (based on (Ca,Mg) O), baddeleyite (based on ZrO_2) and graphite refractories are alkaline. Therefore, they are suitable for 1 applications where the aggressive liquid, slag or atmosphere is alkaline. This is true of furnaces for melting iron and non-ferrous metals, of continuous steel casting equipment and of kilns for production of Portland cement clinker. When an aggressive liquid, slag or atmosphere is acidic,such as a molten glass compositions, acidic materials are used. These include silica (based on SiO_2), aluminosilicate (e.g. based on fireclays or mullite) refractories. Where there are both acidic and alkaline ingredients in the slag and amphoteric is the atmosphere, neutral refractory materials, such as corundum (based on Al_2O_3) or chromite (Cr_2O_3), are used successfully.

Carbide and nitride refractories (based on SiC, Si_3N_4, B_4C or BN) are used mainly for handling hot solid non-ferrous metals (e.g. copper) and in coal-burning furnaces or in waste incinerators. Burning coal at high temperatures produces

Table 2.5 Solubility rate in a basic liquid (FeO)

Material (ranged according to decreasing acidity)	Solubility rate [$mg \cdot cm^{-2} \cdot s^{-1}$]
SiO_2	40.0
Al_2O_3	8.5
MgO	9.4
Refractory chromo-magnesite	4.4

highly aggressive molten alkaline salts and silicates to which resistant are only the carbide- and nitride-based materials. Waste incineration requires (unless a complex system of filters is used) a temperature so high that it can be handled only by the carbide and nitride refractories.

High resistance to thermal shocks and low-thermal conductivity are both ensured by a relatively high porosity of the refractories (Fig. 2.13).

Thermal Stress Sudden temperature changes of the ambient (thermal shocks) cause an inhomogeneous temperature distribution and thus thermal stress in materials. These stresses, especially tensile ones,can bring about brittle fracture. A simple model, shown in Fig. 2.14, will be used to explain this. A very thin plate (with negligible thickness in direction z) is a model of external layers of a hot material, which become cold due to a sudden temperature decrease in the ambient. For a non-zero thermal expansion coefficient α of the material, the plate dimensions tend to decrease on cooling. The model plate, like the external layers of a material, is fixed in direction x to a rigid substrate which is a model of the bulk of the material which does not initially change its temperature and dimensions.

Let us now conduct a mental experiment, in the first stage of which the plate, cooled from T_p to T, freely adapts its dimension to temperature changes in the ambient, so that it shrinks in direction y and x e (direction z can be omitted because

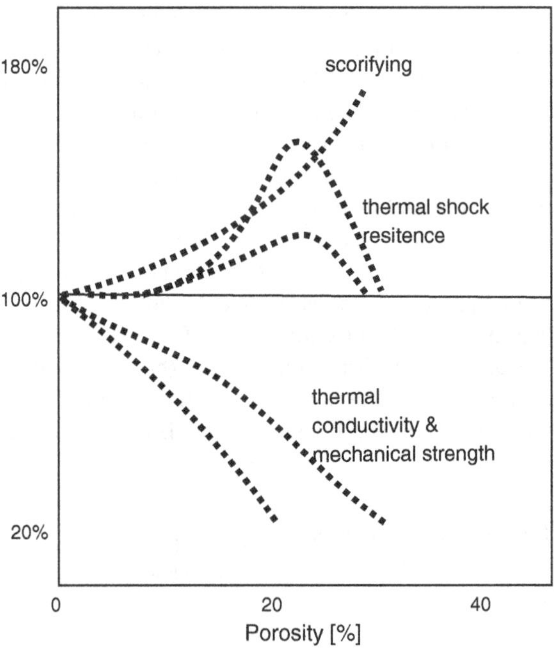

Fig. 2.13 Properties of refractory materials versus porosity. *Note* The properties of non-porous materials serve as a reference point, equal to 100 %

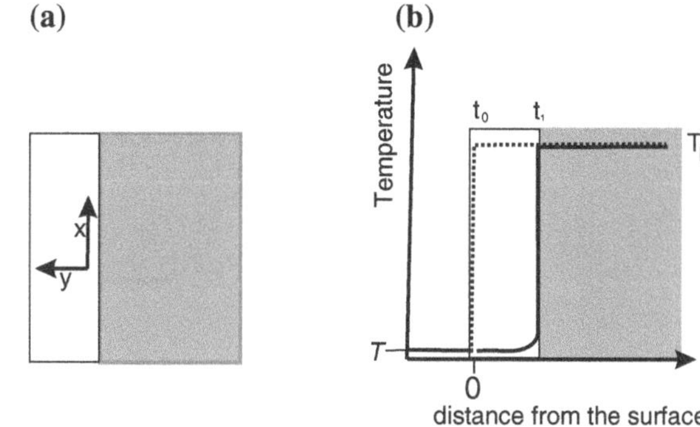

Fig. 2.14 A thin plate rigidly fixed to the substrate which does not change its size withtemperature (**a**). Temperature distribution in the hot model (dotted line) and immediately afterplacing it in a cool environment of a temperature T (continuous line) (**b**)

of the low thickness of the plate). The relative strain in the directions x and y, $\varepsilon_x < 0$, is specified approximately by:

$$\varepsilon_{x=}\varepsilon_y = \bar{\alpha}\left(T - T_p\right) \tag{2.3}$$

where: $\bar{\alpha}$—average linear thermal expansion coefficient for the material of the plate. In fact, the plate can freely change its size in direction y only but not in direction x, because it is rigidly fixed to the substrate. To restore the actual condition, the freely shrinked plate has to be expanded by $\varepsilon_x > 0$, by applying tensile stress $\sigma_x > 0$. Assuming that on sudden dimensional changes the behaviour of the plate is elastic ($\sigma_x = E\varepsilon_x$, where E is the Young's modulus), the Eq. (2.3) van be re-written as:

$$\sigma_x = -E\varepsilon_x = E\bar{\alpha}_x(T_p - T) = E\,\bar{\alpha}\,\Delta T \tag{2.4}$$

The thermally induced stress thus increase with ΔT and at a certain value of ΔT crack propagation can start in the material, reducing the strength of material. It can be assumed that refractory materials, working under the weight of the lining only, can retain their integrity up to 20 % of their original strength. The value of ΔT at which the strength is reduced to this point can be taken as a measure of the material resistance to thermal shocks, ΔT_{max}. As shown in Fig. 2.15, the strength reduction due to thermal stress brought about by sudden temperature changes may take place suddenly or stepwise. The former case is typical for monolithic materials where a once initiated crack propagation continues through the whole material at a speed of sound (Fig. 2.16a). The gradual decrease of mechanical strength due to thermal stresses, found with porous materials, can be attributed to a presence of grain boundaries with various mechanical strength in the porous material. Decohesion occurs locally, first at the weakest interfaces (Fig. 2.16b). As the weak grain boundaries are statistically distributed, a net of separated small cracks is formed in the material which decreases the strength stepwise.

Thermal Conductivity The influence of porosity on thermal conductivity λ [$J \cdot s^{-1} \cdot m^{-1} \cdot deg^{-1}$] shall be discussed by using the mathematical model presented in

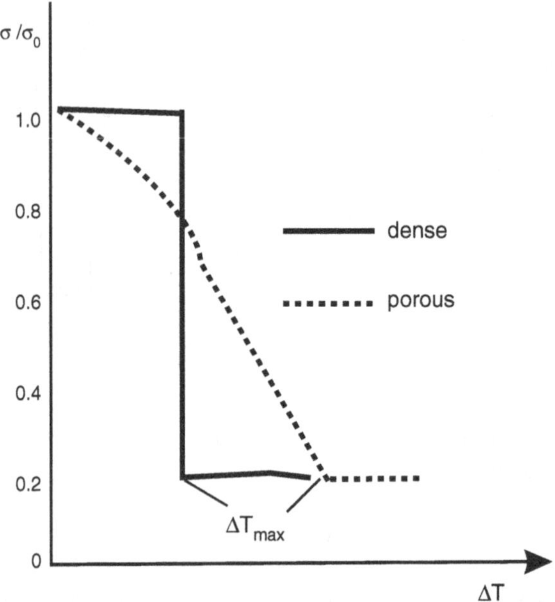

Fig. 2.15 Ratio of bending strength after σ and before σ_o a sudden drop of temperature by ΔT for a dense and a porous material. The maximum temperature difference which the materials can sustain, ΔT_{max}, corresponds here to ΔT at which the ratio decreases to $\sigma/\sigma_o = 0.2$. Typical values of ΔT_{max} range from 200 °C (glass) and 800 °C (porous SiC)

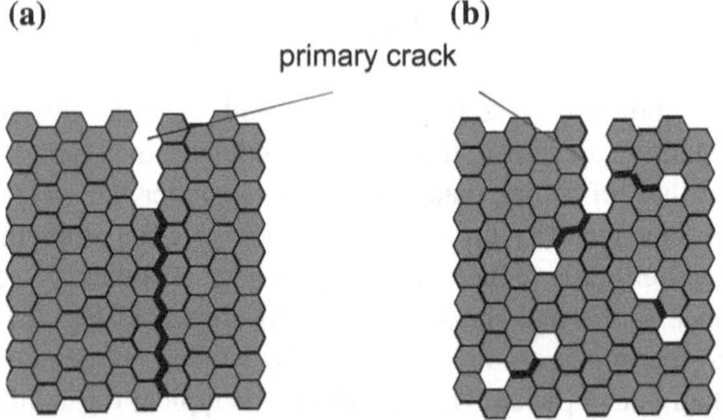

Fig. 2.16 Cracking due to thermal stress: Of a material with homogeneous microstructure **a** and of a porous material with varied mechanical strength of grain boundaries (**b**)

Fig. 2.17. Thermal energy q [J] transported though the model is the sum of energy transported via the phase 1 and 2, respectively:

$$q = q_1 + q_2 \tag{2.5}$$

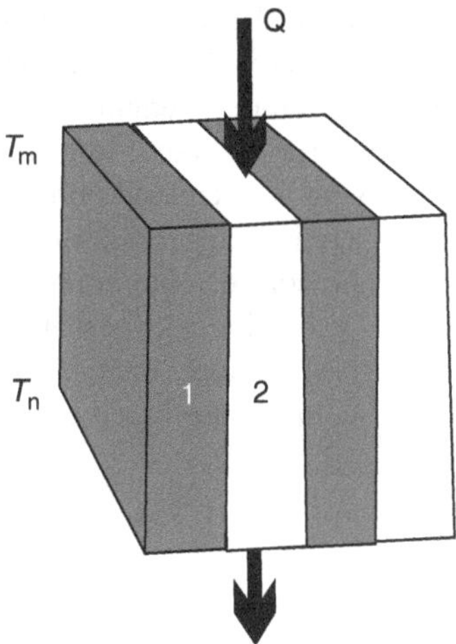

Fig. 2.17 Mathematical model of two-phase materials; phases (marked with indices 1 and 2). In the model: there is a stationary flow of heat (Q) between the surface at temperature T_m and the surface at temperature $T_n < T_m$. $T_m - T_n$ and the temperature gradient are thus equal for both phases

Given the dimension of λ, Eq. (2.5) can be—written as:

$$\lambda(V\Delta Tt/L^2) = \lambda_1(V_1\Delta Tt/L^2) + \lambda_2(V_2\Delta Tt/L^2) \tag{2.6}$$

where the lower indices 1 and 2 stand for the phase 1 and phase 2; V for volume; L for thickness; ΔT for temperature gradient; t—for time. As L, ΔT and t = const, Eq. (2.6) can be rewritten:

$$\lambda V = \lambda_1 V_1 + \lambda_2 V_2 \tag{2.7}$$

Introducing the notion of volume fraction of a phase $V_i' = V_i/V(i = 1, 2)$, Eq. (2.7) yields:

$$\lambda = \lambda_1 V_1' + \lambda_2 V_2' \tag{2.8}$$

Let us assume that index 1 denotes a solid l with better thermal conductivity and index 2 a gaseous phase (contained in the pores) with much lower conductivity $\lambda_2 \ll \lambda_1$. In this case:

$$\lambda = \lambda_1 V_1' = \lambda_1(1 - V_2') \tag{2.9}$$

In other words, heat is conducted mainly by the superior heat conductor 1 (solid), thermal conductivity decreasing linearly with volume fraction of the inferior heat conductor 2, i.e. pores containing gas. Except for fibrous materials, Eq. (2.9) works well for volume fractions of pores up to 20 %.

As shown in Fig. 2.13, mechanical strength of refractory materials decreases along with the volume fraction of pores. In practice, refractory materials perform only under their own weight. Therefore, strength reduction with porosity is less important than the increase of material resistance to thermal shocks and decrease of thermal conductivity accompanying an increased porosity. It must be noted, however, that with higher porosity there is a higher infiltration of the refractory material by an aggressive liquid phase, and considerable infiltration of slag may reduce the life of refractory material. The discussion shows that requirements for refractory materials are often contradictory. Therefore, their phase composition and microstructure are usually a result of compromise based on experience.

Chapter 3
Ceramics to Overcome Brittleness

3.1 Mechanical Properties of Ceramic Materials

Elastic Properties Ceramic materials are used for structural purposes due to such properties as high elasticity and hardness (see Table 3.1), low-specific gravity, and resistance to wear, aggressive atmosphere and high temperatures. These properties are mostly found in materials from synthetically derived crystalline oxides (αAl_2O_3, ZrO_2), carbides (SiC, B_4C) and nitrides (-Si_3N_4 and sialons -Si_3N_4 derivatives). This is related to the chemical bond type in their crystalline structure. They are composed of elements near one another in the periodic table, characterised by similar effective nuclear charges, Z_{ef} (→Chap. 5), of a relatively high value. The bonds have therefore a significant covalent contribution (→Fig. 5.2), making the bonds directional. The spatial polymer structure they form is responsible for the high elasticity, hardness and wear resistance. However, in the single crystalline form they are brittle. If intended for structural purposes less brittle forms are used such as single-phase (monolithic) polycrystalline (Fig. 3.1) or multiphase materials, -composites.

Let us look at the mechanical properties of polycrystalline ceramics and composites in detail. Let $\varepsilon = \Delta l/l_0$ denote relative strain of material (where l_0 stands for the initial length, Δl—for length increase of the material subjected to load), and σ—for load applied to the material. The load evokes in the material a stress, i.e. force acting on a unit of area of a given volume of material. Except for the immediate vicinity of defects of the crystalline structure, the stress has an identical value with the load, σ,

An elastic solid is characterised by a practically linear relationship between relative strain ε, and applied load σ (Fig. 3.2a): $\sigma = C\varepsilon$, where C is an elasticity constant. Moreover, the strain is independent of time and is reversible, i.e. appearing when a load is applied to the material and disappears immediately when the load is removed (Fig. 3.2b).

The elasticity constant often used is the Young's modulus *E:*

R. Pampuch, *An Introduction to Ceramics*,
Lecture Notes in Chemistry 86, DOI 10.1007/978-3-319-10410-2_3

Table 3.1 Vickers hardness of some ceramic materials

Material	Vickers hardness H_v [GPa][a]
diamond	60–137[b]
BN reg	40–50
SiC	27
TiC	26
B_4C	20–30
ZrC	24
αAl_2O_3	20
$Si3N_4$	20
AlN	10–14
TiB_2	13
ZrO_2	12
MgO	7.5
Soda-lime glass	5–10

[a] $H_v = F/A$, where *F*—force applied to a diamond pyramid pressed against the material; *A*—area of impression in material
[b] Value estimated using various methods

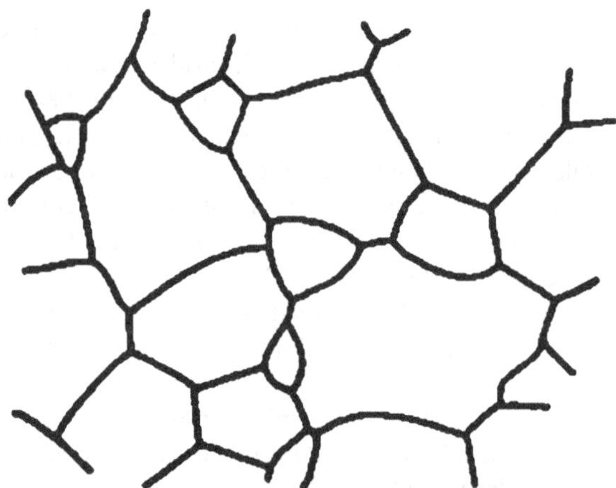

Fig. 3.1 Cross-section of a monolithic polycrystal (schematic)

$$E = \partial\sigma_i/\partial\varepsilon_j(i = j = 1, 2, 3) \tag{3.1}$$

where 1, 2 and 3 stand for coordinates of a rectangular system of space coordinates.

The brittleness of materials manifests itself by brittle fracture, a fracture of the material into two or more parts which is due to spontaneous propagation of cracks of a given critical size. The propagation, once initiated, occurs at a very rapid rate (2–5 km·s^{-1}) at a low relative strain ε_c.

Brittle Fracture Methods for estimating the risk of brittle fracture originate in concepts proposed by Griffith. To this end, let us assume that the material is in the

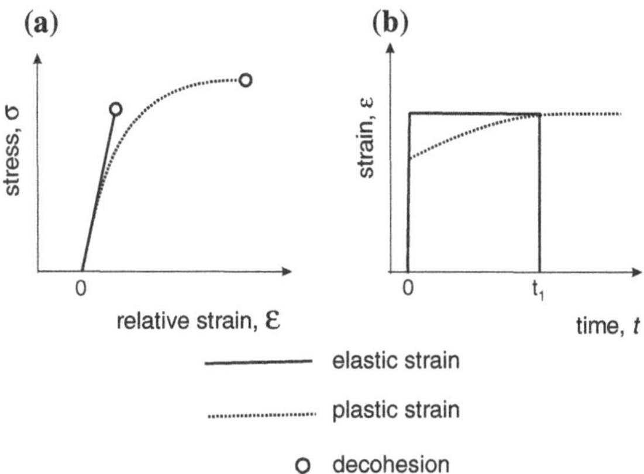

Fig. 3.2 Relative strain ε of elastic material versus stress σ **a** and relative strain of a loaded elastic material versus time *t* (**b**). *Note* data for a Bingham body which undergoes plastic strain is given for comparison (*dotted line*)

form of an infinitely thin, plane-stressed plate, meaning that the external load performed work which had been stored in the material as elastic strain (potential) energy. It can be written as

$$U = 2\sigma\int_0^{\varepsilon} d\varepsilon = 2\sigma\varepsilon = 2(\sigma^2/\mathrm{E}) \tag{3.2}$$

where: *U*—stored elastic strain energy density [$\mathrm{J\cdot m^{-3}} \equiv \mathrm{N{\bullet}m^{-2}}$]; σ—stress; ε—relative strain; *E*—Young's modulus. If a crack of length *a* is introduced into the thin plate it causes the material stress to relax and elastic stress energy to be released along this length, approximately in the shaded area in Fig. 3.3, equal $A = (1/2)\pi a^2$. Therefore, the energy released, U_R [$\mathrm{J\cdot m^{-1}}$], per unit crack length is given as

$$U_R \approx U \cdot A = (\sigma^2/E)\pi a^2 \tag{3.3a}$$

Advance of a crack by unit length dissipates energy owing to creation of new gas–solid interface with surface energy γ

$$U_D = \gamma\, a \tag{3.3b}$$

The crack can propagate spontaneously when during its increase in size $U_R \geq U_D$, that is, when the production of energy becomes equal to or greater than its dispersion, i.e. when:

$$(\sigma_c^2/E)\pi a_c \geq \gamma \tag{3.4}$$

The left-hand term is called the critical strain energy release rate G_c, the right-hand one—material's resistance (to crack propagation). Materials resistance

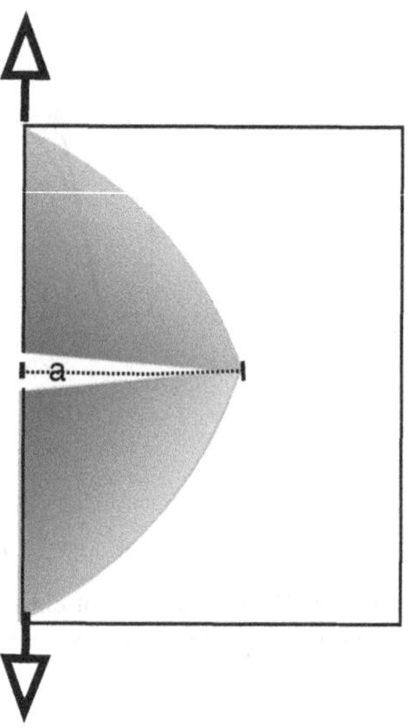

Fig. 3.3 Griffith's model: crack with a length of *a*, formed in a very thin plane-stressed plate causes release of stored elastic strain energy in the darkened area

R have been earlier identified with surface energy γ of the new crack surfaces formed during propagation (Fig. 3.4a) but it is advisable to treat it as a complex quantity, including other energy sinks in the crack environment:

$$R = \gamma + \gamma_{\text{microcracks}+}\gamma_{\text{plast}}\cdots\gamma_{i} \tag{3.5}$$

indicating that the elastic strain energy can be dispersed also by formation of microcracks, plastic flow and other phenomena accompanying crack propagation (Fig. 3.4b). In the latter case, $G = G_c$ when:

$$\left.\frac{\partial G}{\partial a}\right|_{\sigma=\text{const}} = \mathrm{d}R/\mathrm{d}a \tag{3.6}$$

Stress Intensity Factor and Fracture Toughness Given the relatively simple experimental determination, a factor called stress intensity factor *K* is used more often than *G*. It is closely related to *G*. In the case of a thin plate where thickness can be neglected, the relation is given as

$$K = (GE)^{1/2} = \sigma(\pi a)^{1/2} \tag{3.7}$$

where: *E* stands for the Young's modulus. The stress intensity factor is usually determined by loading the material in the mode *I*, shown in Fig. 3.5. In this mode triaxial stresses are evoked, creating the highest risk of crack propagation. Its critical value is called: fracture toughness, K_{Ic}:

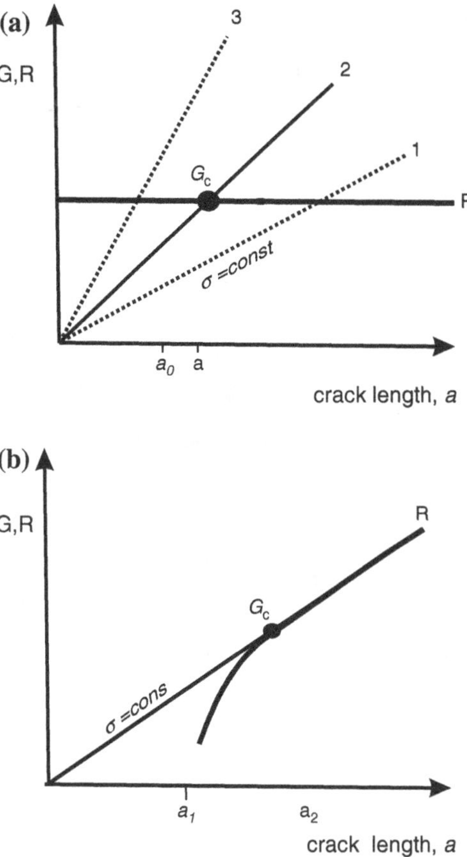

Fig. 3.4 Elastic strain energy release rate *G* versus crack size *a* for a constant (Fig. 3.4a) and *a* variable resistance *R* of the material (Fig. 3.4b). According to Eq. (3.4), in the coordinate system: y = *G, R; x* = *a*, there is a linear relationship between G and *a*, $G = (\sigma^2/E)\pi \cdot a$, where: *σ*—stress; *E* —Young's modulus. *Note* for stress *σ* corresponding to the straight line 1, the stress release rate *G* on an increase of crack length from a_0- to *a* is always less than material resistance *R*, so that no spontaneous increase of the crack size can occur. $G = G_c \geq R$ can be, however, attained on crack size increase from a_o to a when the stress is increased to a value corresponding to the straight line 2; the crack can thus propagate spontaneously. The case of rising *R* (Fig. 3.4b) is explained in detail in the text

$$K_{Ic} = \sigma_c(\pi a_c)^{1/2} \tag{3.8}$$

where: a_c and σ_c denote, respectively, the critical crack length and stress value at which the crack begins to propagate spontaneously. To include the effect of crack location and its shape, a full formula for K_{Ic} includes also an additional constant Y, close to 1. According to extensive collection of data, Eq. (3.8) is fulfilled by most monolithic polycrystalline ceramic materials and nearly theoretically dense

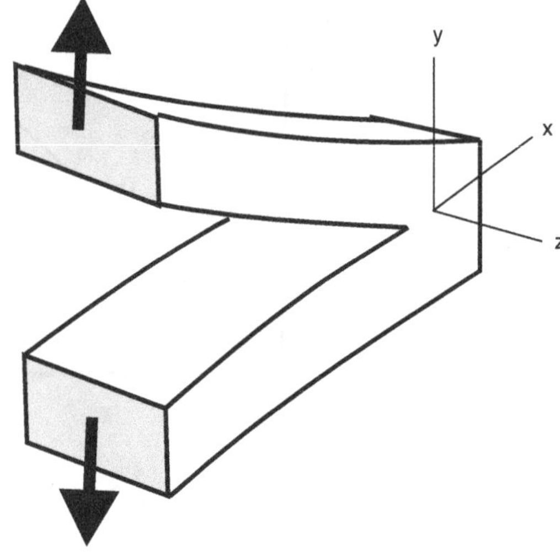

Fig. 3.5 Opening of *a* crack in the plane-deformed state (mode *I* of fracture). There is a three-axial stress in the mode *I* and the highest risk of brittle fracture

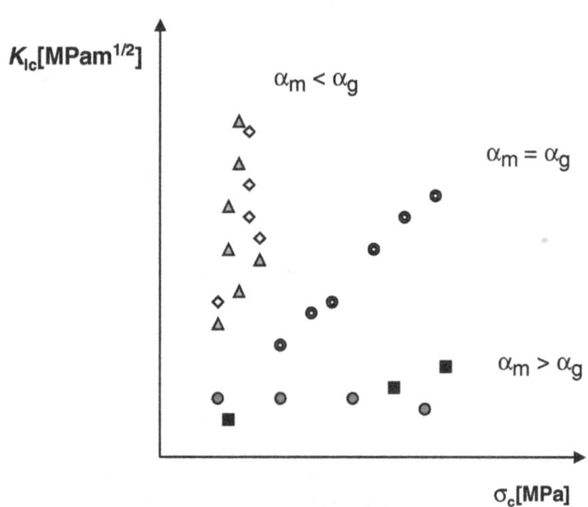

Fig. 3.6 Fracture toughness K_{Ic}, versus tensile strength σ_c, for found with particulate composites. *Note* the data are for nearly theoretically dense composites, produced in a single laboratory by using identical processing; in this case variations of secondary factors have a minor influence on the results

particulate composites, provided their porosity and particle size is similar (Fig. 3.6). Therefore, it can be considered as the basic equation of linear fracture mechanics.

3.2 Mechanical Properties of Polycrystals and Particulate Composites

Composites are materials made of at least two solid components with deliberately selected properties. After joining them into one material, a synergic effect can be observed, i.e. obtaining properties superior to those of the individual components. This section focusses primarily on the potential for achieving higher fracture toughness and higher mechanical strength in the composites.

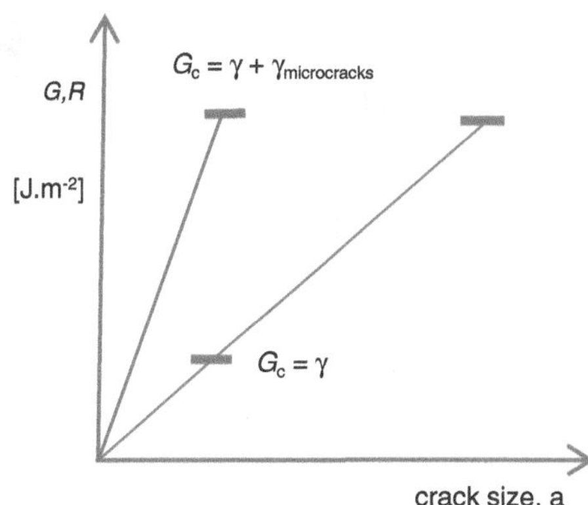

Fig. 3.7 Effects of increasing *R*

One type of such materials are particulate composites, consisting of a polycrystalline matrix in which are dispersed micrometre-sized or sometimes nanometre-sized particles of a second solid phase. The processing used to produce the composites is common to most kinds of ceramics, i.e. shaping and sintering of powder mixes (→Fig. 1.7).

Polycrystals and Particulate Composites vs. the Basic Equation of Linear Fracture Mechanics Up to a certain volume fraction of dispersed particles, V'_g, a substantial increase in mechanical strength σ_c and/or fracture toughness K_{Ic} can be attained here, and it is of interest to consider the factors which play here a role. The discussion is limited to polycrystals and particulate composites where variations of porosity and grain size influence their inherent properties to a minor degree, and these materials fulfil the basic equation of linear fracture mechanics (3.8).

The obvious way to rise G_c (and K_{Ic}) is an increase of *R*. An influence of plastic flow on *R* can be excluded in ceramics because the number of independent glide systems is here lower than five. However, *R* can be increased by energy sinks in the form of metastable microcracks. The effects are graphically illustrated in Fig. 3.7 using a *G*-*a* graph (see, also Fig. 3.4) with coordinates $x = a$; $y = G = m \cdot a$; $m = (\sigma^2/E)\pi$.

Role of Residual Thermal Stress: Higher G_c (and K_{Ic}) can be achieved in limiting cases either by an increase of the critical stress (σ_c) or an increase of the critical crack size a_c.The former case is commonly found with polycrystals and particulate composites where the matrix and the dispersed particles have similar Young's moduli *E* and thermal expansion coefficients α. The latter case is typical for particulate composites where the dispersed particles have a higher thermal expansion coefficient than the matrix $\alpha_g > \alpha_m$. Let us consider this case in detail. Figure 3.8 shows that hydrostatic stress at the matrix-particle interfaces which arise on cooling down sintered composites should be compressive ($p < 0$) in the case of $\alpha_m > \alpha_g$ and tensile ($p > 0$) in the case of of $\alpha_m < \alpha_g$—; *m* stands for the matrix and *g* for the particles. The resulting stress distribution is schematically shown in Fig. 3.9. The

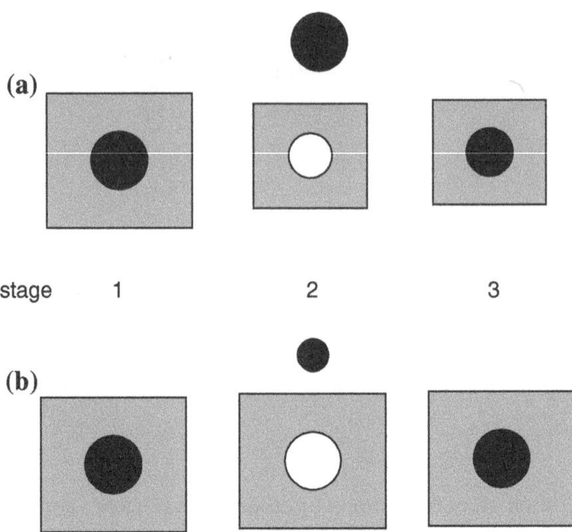

Fig. 3.8 Mental experiment illustrating size changes while cooling separated particles (g) and matrix (m) of a composite from a high sintering temperature to room temperature; to restore the real situation the separated parts are recombined, matching up different shrinkage in previous stage. Note: the upper schema relates to the case $\alpha_m > \alpha_g$, resulting in a compressive residual hydrostatic stress at the matrix–particle interface; the lower schema relates to the case $\alpha_m < \alpha_g$, resulting in a tensile hydrostatic stress at the interface

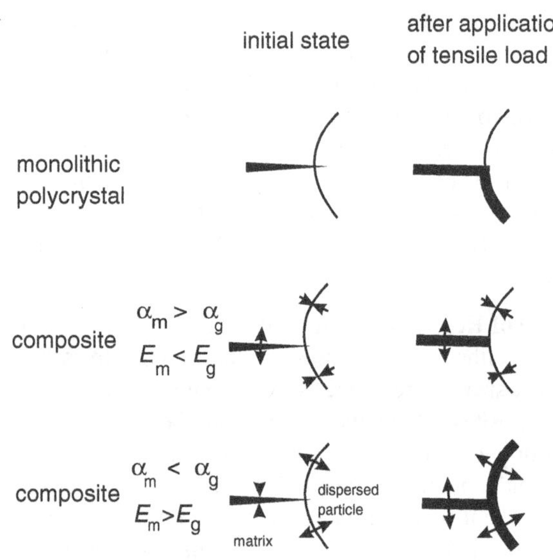

Fig. 3.9 Tensile and compressive stresses at interfaces in a monolithic polycrystal and a particulate composite (schematic). After application of tensile load, the distribution of stresses should bring about crack bridging in the case of $\alpha_m > \alpha_g$ and crack branching in the case of $\alpha_m > \alpha_g$—, respectively

Table 3.2 Sample properties of particulate composites

Phase composition matrix-dispersed particles	$\alpha_m < \alpha_g$	$\alpha_m \approx \alpha_g$	$\alpha_m > \alpha_g$	Increase of K_{Ic} with V'_z; σ_c = const	Increase of K_{ic} and σ_c with V'_z	Increase of σ_c; with $V'z$; K_{ic} = const
SiC-TiB_2	+			+		
SiC-B_4C	+			+		
SiC-B_4C + TiB_2	+			+		
AlN-TiB_2	+			+		
Sialon-SiC		+			+	
Al_2O_3-ZrO_2		+			+	
Al_2O_3-SiC			+			+
SiC-C (graphite)			+			+

Note α_m and α_g represent the thermal expansion coefficient of the matrix and dispersed particles, respectively; $V'z$—volume share of particles; K_{Ic}—fracture toughness; σ_c—bending strength

case $\alpha_m < \alpha_g$ (→lowest schema in Fig. 3.9) creates microcracks around the dispersed particles. They are thus sites of decreased elastic strain energy, which can be linked together via cracks along paths of reduced elastic strain energy constituted by grain boundaries in the matrix. In this way cracks can extend to appreciable lengths without achieving a critical size.

The line of reasoning adapted explains rationally experimental data for nearly theoretically dense composites collected in Table 3.2. While the combinations $\alpha_m > \alpha_g$ and $\alpha_n \approx \alpha_g$ result in increased critical strength, the combinations $\alpha_m < \alpha_g$— show an increased fracture toughness linked with small changes of the critical stress only.

Controlled microcracking is operative also in special cases, such as ZrO_2 s.s stress-induced transformations and in polycrystalline ceramics matrix composites reinforced by ceramic fibres where the fibre–matrix interfaces have been made deliberately weak. When a crack propagating in the polycrystalline matrix reaches a weak interface with the fibres, it branches along that interface (stage 2 in Fig. 3.10). The branched crack is subjected to weaker stress than the initial crack. As a result of branching, a net of metastable microcracks is formed. Fibres not yet damaged at this stage continue to bridge the matrix cracks (stage 3 in Fig. 3.10). However, the issue is more complex. The fibre–matrix interfaces have to be not only weak enough to promote the branching of cracks, but also strong enough to ensure the transfer of stress from the matrix to the stronger fibres. These and other, often contradictory,

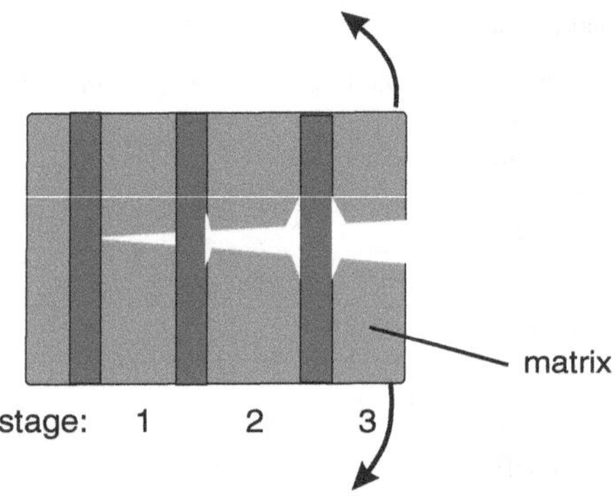

Fig. 3.10 Successive stages of cracking (1, 2, 3) of a fibre-reinforced composite with a weak fibre–matrix interface

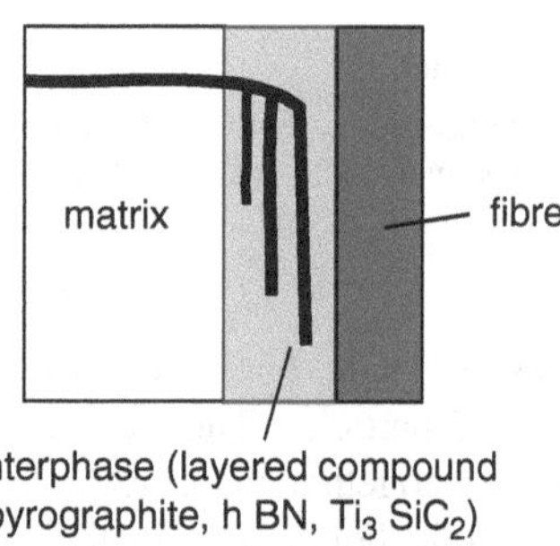

Fig. 3.11 Cracking of an intermediate layer in a ceramic matrix composite reinforced with ceramic fibres

requirements can be met by introducing, between the fibres and matrix, a layer of an intermediate substance, often characterised by multi-layer crystalline structures such as graphite, hexagonal BN or Ti_3SiC_2. Instead of a simple delamination at the weak fibre–matrix interface, multiple delaminations can be observed in the intermediate layer (Figs. 3.10 and 3.11).

3.3 Use of Ceramic Materials in Fibre-Reinforced Polymer Composites

The first realised kind of fibre-reinforced composites was in form of preforms where fibres, unidirectionally aligned or weaved or processed in one or another way, were impregnated by epoxide, phenol formaldehyde or imide resins. As mechanical strength of materials in fibrous form is usually one order of magnitude higher than

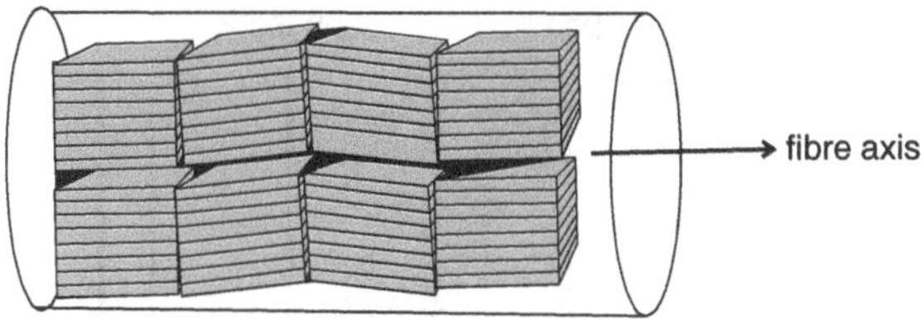

Fig. 3.12 A model of high-modulus carbon fibres, composed of ribbons of packets of hexagonal carbon atom layers (→Fig. 5.8), oriented nearly parallel to the fibre axis. Continuous carbon fibres have exceptional mechanical properties: Young's modulus up to 830 GPa and tensile strength up to 5 GPa, higher than steel

in polycrystals, the polymer enveloping the fibres serves to transfer the load to the load-bearing fibres.

Several types of continuous ceramic fibres are used here. Fibres made of *E*-type glass s (→2. Tradition continued. The silica tradition) in composites used for construction of boats, carbon fibres (Fig. 3.12) in composites used for aviation and sports equipment. Materials other than the continuous fibres can also be applied. Namely, particles where at least one dimension is <100 nm, like clay minerals (→2. The tradition continued. The clay tradition), carbon nanotubes and graphenes (→5. Unusual ceramic dielectrics and conductors). Due to a high ratio of length to thickness such fillers are in contact with the polymer matrix over a large area. Therefore, their effect on strength and fracture toughness—can be observed even with a low-volume fractions of fillers (from 0.5 to 15 %wt). Due to the dimensions of these fillers, these composites are often referred to as nanocomposites.

3.4 Functionally Graded Composites

Some applications require composites in which phases of radically different properties have to be joined, such as ceramics with metals or rigid hydroxyapatite with highly deformable collagen. Such applications include the construction of space vehicles (with a ceramic layer outside and a metallic structure inside a vehicle), biocomposites, gas turbine blades, machining tools, attack vehicle armour and many others.

However, in the case of such components, there is a risk that excessively high residual thermal stress occurs, due to the very different thermal expansion coefficients, thermal conductivity, Young's modulus, etc., of the component phases.

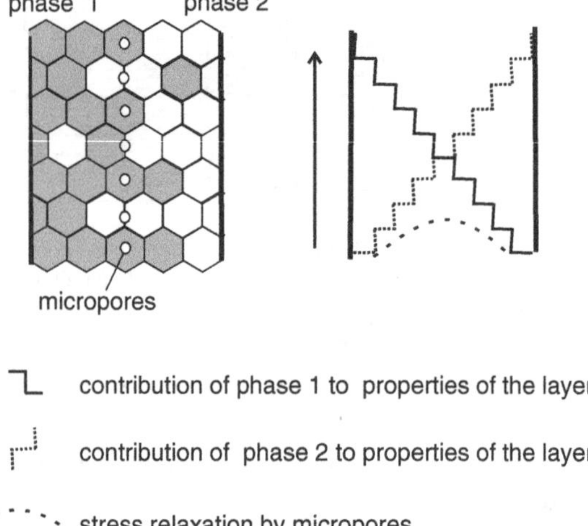

Fig. 3.13 Changes of composition and in properties of a functionally graded material (FGM)

To avoid this, the composites consist of layers, with thicknesses in micrometers or millimetres, having a composition which is modified in stages (Fig. 3.13). This results in nearly continuous changes in properties throughout the composite. These composites are known by the acronym FGM, standing for functionally graded materials.

Chapter 4
Voids Are Important

An important class of ceramic composites are materials in which voids (pores) of various shapes and volumes are created in a controlled manner. In these materials, the properties of these empty spaces are combined with those of the ceramic solid phase. There is also a great potential in materials in which the solid phase is just a framework for liquids or gases performing important functions.

Such materials include primarily the following: 1. cellular materials containing tubular pores with small diameters within a range of 10 nm to more than 20 μm (Fig. 4.1a); 2. foam materials with a high-volume fraction (0.7 or more) of partially connected isometric pores (Fig. 4.1b); these also include aerogels, derived from gels, in which the liquid component is replaced with gas; 3. molecular sieves in which (usually interconnected) voids with sizes of fractions of a nanometre appear.

4.1 Cellular Materials

Catalytic Converters The classic materials of this type are exhaust gas catalytic converters, made by extruding a plastic cordierite ($2MgO \cdot 2Al_2O_3 \cdot 5SiO_2$) paste through a nozzle (→Fig. 1.3). Walls of tubular pores with diameters of 0.1–0.2 mm are covered in the converter with layers of porous γ Al_2O_3 containing dispersed Pt and PtRh catalysts. On the surface of the tubular-pore walls (with a surface area up to 25,000 m^2), catalytic oxidation of hydrocarbons, CO to CO_2 and H_2O, and reduction of NO_{x-} to N_2 occur in the flowing exhaust gases.

Wettability and Non-wettability of Surfaces An equally broad range of applications have membranes and filters made of cellular ceramic materials. One of the features of ceramics, indispensable in such applications, is the ability to suck liquids through tubular pores. This requires wetting of the solid surface by a liquid. Wettability and non-wettability can be assessed by considering the equilibrium conditions in the solid–liquid–gas system:

R. Pampuch, *An Introduction to Ceramics*,
Lecture Notes in Chemistry 86, DOI 10.1007/978-3-319-10410-2_4

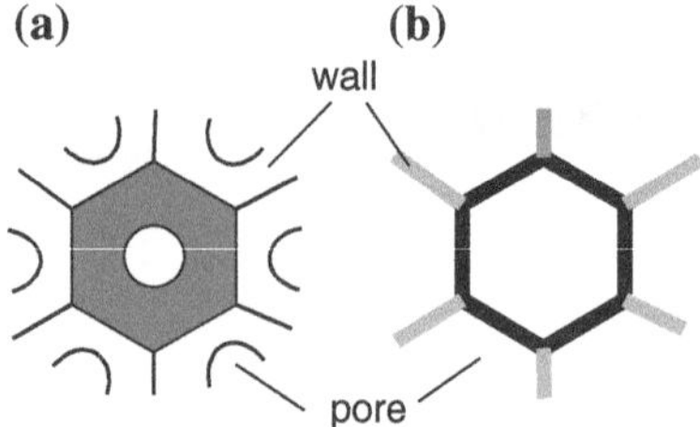

Fig. 4.1 Cellular (**a**) and foam material (**b**)

$$\gamma_{sg} = \gamma_{sc} + \gamma_{cg}\cos\phi \quad (4.1)$$

where: γ_{sg}—surface tension at the solid-gas interface; γ_{sc}—surface tension at the solid–liquid interface; γ_{cg}—surface tension at the liquid–gas interface; ϕ—contact angle. For $\phi < 90°$ there is a wettability of the solid (Fig. 4.2a), and for $\phi > 90°$ the liquid does not wet the solid (Fig. 4.2b). According to equation (4.1), the condition $\phi < 90°$ can be met when $\gamma_{sg} - \gamma_{sc} > 0$, i.e. when the surface tension at the solid–gas interface, γ_{sg}, is substantial In contrast to, for example, organic polymers, ceramic materials are distinguished by high values of γ_{sg} and, therefore, are usually well wetted by liquids.

Filters and Membranes Given a narrow range of tubular pore diameters, the flow of liquid or gaseous suspensions of particles or molecules via the pores allows permeation only of those particles or molecules that are smaller than the pore diameter. Cellular materials can therefore act as a filter which stops undesired particles, making the secondary stream of fluid cleaner (Fig. 4.3). Using this effect it is also feasible to fulfil the opposite function. Namely, through sucking liquid from suspension, valuable molecules (e.g. proteins) previously dispersed in liquid can be concentrated. Materials with cellular pores of a diameter $d < 10$ μm are usually called *membranes*, whereas the term *filters* is reserved for materials with cellular pores of larger diameters. Typical diameters of cellular pores and applications of ceramic filters and membranes are listed in Table 4.1.

Figure 4.3 indicates that ceramic membranes usually comprise multiple layers, composed of a porous substrate (usually Al_2O_3) containing pores with diameters of 1–10 μm and intermediate and upper layers in which fluids and solid particles are

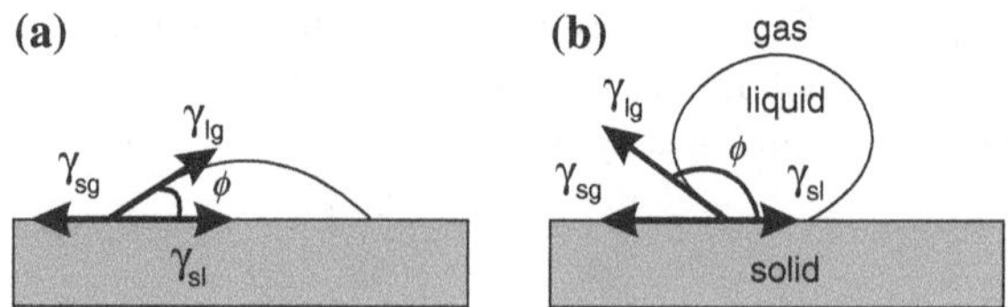

Fig. 4.2 Wetting (**a**) and non-wetting (**b**) of a solid surface by liquid

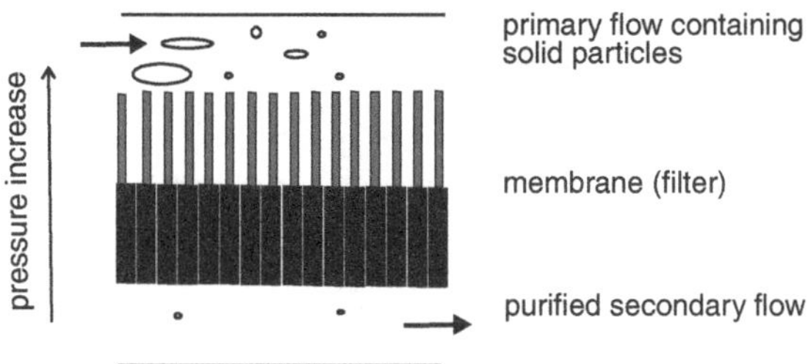

Fig. 4.3 Operating principle of ceramic filters and membranes

Table 4.1 Ceramic membranes and filters and their applications

Type	Pore diameter (μm)	Applications	Typical material
Molecular sieves	0.0002–0.0003	Selective absorption of H_2, CO, CO_2, NH_3	Dehydrated zeolites
Membranes	0.001–0.1	Ultrafiltration: protein separation, concentration and blood fractionation, blood filtering	Mesoporous Al_2O_3, ZrO_2, SiO_2
Membranes	<10	Microfiltration of beverages (beer, wine), densification of skimmed milk, removing oil from water	Al_2O_3, ZrO_2
Filters	>10	Separation of non-metal particles from molten metals, separation of dust particles from industrial and flue gases	SiC, mullite ($3Al_2O_3 \cdot 2SiO_2$), cordierite ($2MgO \cdot 2Al_2O_3 \cdot SiO_2$)

separated. The pores of the upper layers are characterised by much smaller diameters (for example, between 2 and 200 nm). To achieve this, specific production methods are required, such as anodisation of thin layers of metallic aluminium or the sol–gel method in which mesoporous TiO_2 or γAl_2O_3 are obtained. One factor promoting the use of porous ceramic materials as filters or membranes is their chemical resistance and retention of form at high temperatures. These qualities have filters made from SiC, Al_2O_3 and ZrO_2, which are used to remove solids from hot aggressive gases and liquid metal alloys. In the food and biotechnology industries, the chemical resistance of ceramic materials to a broad range of cleaning liquids are also important. For example, liquids with pH values from 2 (hydrochloric acid) to 10 (aqueous solution of calcium hydroxide) are neutral to Al_2O_3 and ZrO_2. This enables the cleaning of membranes and filters in acids or bases.

Microfluidics Cellular materials with tubular pores play an important role in microfluidics, i.e. a technology which works with small volumes of liquids. Fluids behave in a specific manner when flowing through channels (tubular pores) with a diameter of a few microns. The relationship between gravity and frictional viscosity forces in liquid flow is expressed by a dimensionless Reynolds number:

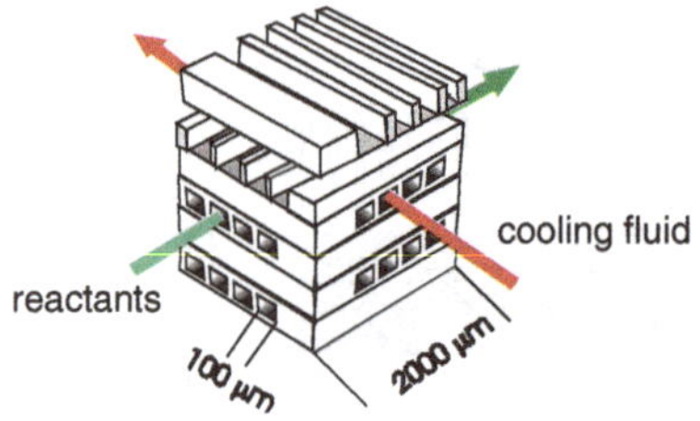

Fig. 4.4 A ceramic micro-reactor

$$[\text{gravity force}/\text{viscosity force}] = \text{Re} = \bar{v}d/\nu \tag{4.2}$$

where: $\bar{v}$ is the average flow rate in relation to the channel walls [m s^{-1}]; *d*—channel diameter [m]; ν—kinematic viscosity of liquid [m^2 s^{-1}]. When the force of gravity dominates, then the flow of liquid is turbulent. With small channel diameters, however, the forces of friction dominate, and the flow becomes laminar, i.e. all liquid layers move parallel to the channel walls. In other words, transfer of mass occurs only in the direction of liquid flow along the channel axis. As experience shows, laminar flow dominates at values of Re < 2,000. For typical kinematic viscosities of liquids: $1-10 \times 10^{-6}$ [m^2 s^{-1}] and flow rates of $1-10$ m s^{-1}, this state can be achieved when the diameter of tubular pores is around 100 µm (1×10^{-4} m).

Laminar and Turbulent Flow of Fluids Rate of laminar flow rate is proportional to the hydraulic pressure gradient and, in contrast to turbulent flow, considerable flow rates can be achieved at low pressure gradients. At the same time, the liquid stream has a high surface to volume ratio; the ratio can reach 20,000 m^2/m^3. This ensures a very high degree of heat exchange between the hot liquid stream and the environment.

Microreactors Thanks to such features, it has been possible to develop a series of unique devices. High flow rates of liquids at low pressure gradients are necessary in the case of inkjet printer heads. This property, combined with intense heat exchange with ambient air, has been the reason for the development of microreactors, i.e. miniature reactors with dimensions that often do not exceed those of a postage stamp. A microreactor is shown schematically in Fig. 4.4. Intense heat exchange with ambient air in microreactors can be deliberately used to safely perform highly exothermic reactions. In the case of conventional large reactors, with large volumes of reactive liquids or gases, it is difficult to dissipate heat and often the risk of explosion is involved. Various materials are used to make microreactors. Given their small dimensions, they can be successfully made of glass, SiC or αAl_2O_3, enabling full exploitation of the unique chemical and thermal resistance of ceramic materials. This is particularly useful for reactions with highly reactive reagents, such as the ones used in splitting water into hydrogen and oxygen by the iodine-sulphur process. The process requires temperatures of around 1,100 °C and use of sulphuric acid. Thanks to high flow rates and efficient heat exchange, the productivity of microreactors is surprisingly high: a cluster of 20–25 interconnected microreactors enables production of chemicals at the rate of 1×10^3 tons per annum.

Lab-on-a-Chip An interesting application has also been found for ceramic materials with micro-tubular pores for production of labs-on-a-chip, used to draw and analyse small samples of liquids and gases which can be used for quick analysis of soil and water contamination, or for immunological tests to detect infections or diseases. An advantage of making labs-on-a-chip from ceramic materials is that, for example, in αAl_2O_3, there are no active centres in which ion exchange may occur. Such an exchange occurs in the case of zeolites and many polymers which, of course, distorts the results of the analyses.

4.2 Foam Materials

Mechanical Properties of Foam Materials Most applications of this type of materials were undergoing tests as this book was being written. Nevertheless, they are worth mentioning. First, practically all types of ceramic substances can be made in this form. Second, a large controlled volume fraction of pores, often interconnected, enables the combination of the unique properties of ceramic materials with a large surface area. Moreover, foam materials have unique mechanical properties. The lesson here is given by nature, which uses porous materials to build large structures (trees, reefs, human bones) which are more flexible than manmade materials. Figure 4.5 shows a typical stress σ versus strain ε graph of a foam material. At first, the deformation is elastic and in accord with the equation $\varepsilon = E_s/\sigma$, where E_s is Young's modulus of the solid phase forming foam-material walls. The following stage, however, is characterised by a relatively high increase in strain under a slowly increasing stress, being the result of flexure of the lattice structure of the foam material.

Therefore, to deform the foam material at this stage, considerable energy, represented by the area under the σ versus ε curve in the stress–strain graph in Fig. 4.5, must be used. Collapse of the foam material due to flexure of the walls finally leads to their compaction. Young's modulus therefore increases, and further increases in deformation now require a considerable increase in stress leading to fracture.

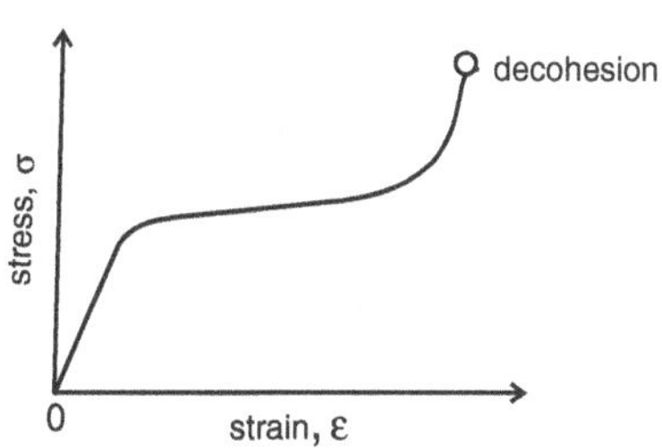

Fig. 4.5 Stress σ-strain ε graph for a foam material

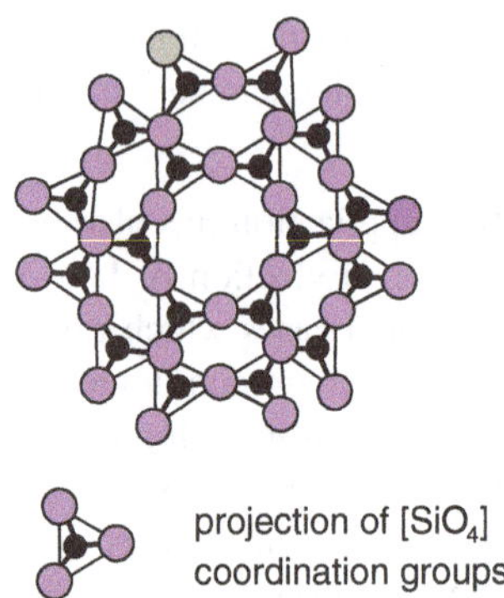

Fig. 4.6 Planar projection of a three-dimensional frame of $[SiO_4]$ coordination groups; if Si atoms are substituted by Ali in some groups it may represent the structure on a dehydrated zeolite

All of this creates numerous potential areas for the application of ceramic foam materials. These include applications in which the controlled flow of liquids through materials is important substrates for catalysts, filters for hot gases and liquid metals, scaffolding used for tissue engineering(→ 7. Imitating and supporting nature). Other applications are absorbing impact energy and high-temperature thermal insulating.

Carbon Nanofoam Products Unique properties characterise carbon nanofoams whose walls feature a well-developed graphite-like structure (→Fig. 5.8). Along with the low density common to all foam products, they offer relatively high mechanical strength. This is due to both the foam structure and the high strength of C–C bonds (→ 3. Ceramic to overcome brittleness and Fig. 3.12). Their unique characteristic is, however, a very high thermal conductivity (up to 180 W m^{-1} K^{-1}) and magnetic properties. The carbon nanofoams are attracted by solid magnets and below −183 °C become magnetic themselves. These properties have led to predictions regarding the use of carbon nanofoams for heat dissipation and cooling of high-power electronic equipment, and for making shields to prevent penetration of electromagnetic radiation.

4.3 Molecular Sieves

The term 'molecular sieve' refers to the fact that their empty spaces are so small that they allow the passage of small molecules only, such as CH_4, CO, N_2 or CO_2. Typical molecular sieves are dehydrated zeolites, aluminosilicates whose structure (with around 130 variations) is formed by a three-dimensional framework of $[SiO_4]$ and $[AlO_4]$, coordination tetrahedra, interconnected by oxygen bridges. In this spatial structure, there are interconnected cages and channels with diameters from around 0.3 to 1 nm (Fig. 4.6), either empty or containing weakly bound cations,

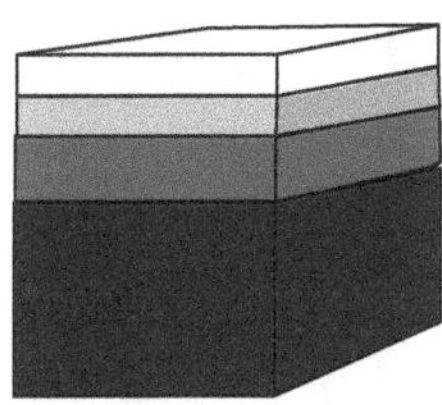

Fig. 4.7 Molecular sieve formed of a series of layers made of several synthetic types of aluminium oxide

such as Na^+ or K^+. Another type of molecular sieve, this one synthetic, composed of multiple layers of aluminium oxide with tubular pores of different diameters, is shown in Fig. 4.7.

Chapter 5
Unusual Ceramic Dielectrics and Conductors

5.1 Dielectrics

Chemical Bond Types in Dielectrics From the point of view of interaction with external electromagnetic (EM) fields, ceramic materials are either semiconductors, the valence electrons of which interact with near and remote EM fields (→ 6. Materials vs. light), and dielectrics where the interaction is restricted to near external EM fields. The near external EM fields are able to shift electric charges in the dielectrics (Fig. 5.1), i.e. polarizing the material. The relationship between material polarization $\overline{P}$ [$C \cdot m^{-2}$] and external field $\overline{E}$ [$V \cdot m^{-1}$] is linear:

$$\overline{P} = \varepsilon_0 \chi_e \overline{E}, \tag{5.1}$$

where: ε_0—dielectric permittivity of a vacuum ($8.854 \cdot 10^{-12}$ [$F \cdot m^{-1}$]); χ_e—dimensionless constant, electrical susceptibility of the material (in technical literature, the relative dielectric constant $\varepsilon_r = 1 + \chi$ is more often used).

Due to Polarization an internal electric field is formed, opposite to the external field's direction. This reduces the total field in the material, enabling storage of electric energy. Therefore, ceramic dielectrics with significant values of χ (or εr) have been used as capacitors in electric circuits for a relatively long time.

In this chapter, however, are considered pyroelectric, ferroelectric and piezoelectric effects owing to which the dielectrics are increasingly exploited in recent years.

Ionic Dielectrics The frequency of near external EM fields used here is typically lower than 10^{11} [Hz] and substantial contribution to Polarization comes from ionic Polarization. The strongest effects are thus observed with ionic dielectrics which are built by cations and anions. Formation of an ionic material is most probable when the effective (i.e. screened by internal orbitals) nuclear charges Z_{eff} of the bond-forming elements strongly differ. This favours a transfer of valence electrons from the element becoming cation to the element becoming negatively charged anion

R. Pampuch, *An Introduction to Ceramics*,
Lecture Notes in Chemistry 86, DOI 10.1007/978-3-319-10410-2_5

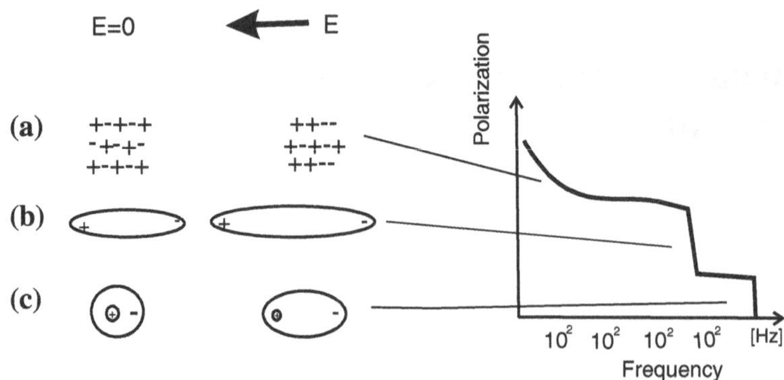

Fig. 5.1 Electric polarization in ceramic dielectrics versus frequency of EM fields. *Note* **a** space charge polarization; **b** ionic polarization; **c** electronic polarization

(Fig. 5.2a). To typical ionic dielectrics belong complex oxides with perovskite structure (Fig. 5.3a, c), such as $BaTiO_3$, PZT, i.e. [$Pb(Zr_yTi1_{-y})O_3$], or PZLT (PZT doped with La ions). At high temperatures, they have a cubic structure which is transformed, on cooling below the so-called Curie temperature, into a structure with lower symmetry. For instance, in $BaTiO_3$, a tetragonal structure, having a single symmetry axis, is formed. This is accompanied by formation of dipoles by a slight displacement of Ti^{4+} cations from a central position along the [001] axis (Fig. 5.3b). The dipoles strongly mutually interact, which increases the electric susceptibility to

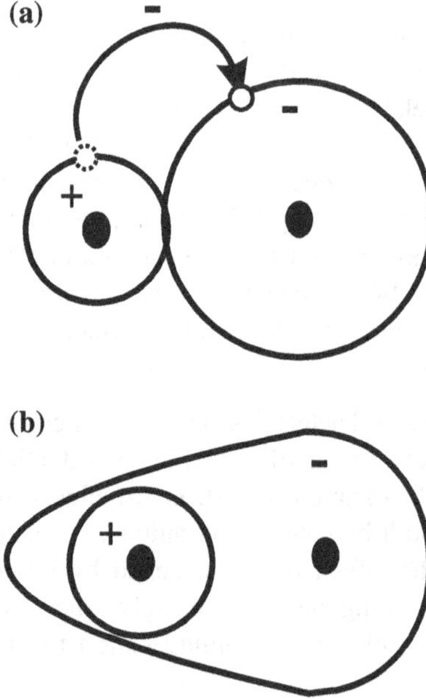

Fig. 5.2 Formation of a ionic bond by transfer of an s-orbital valence electron, giving rise to cation, anion and ionic bond (**a**); sharing of s-orbital valence electrons in a partly polarised covalent bond (**b**)

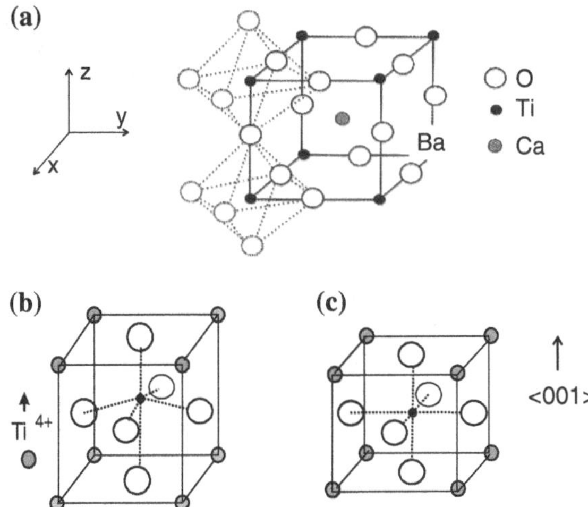

Fig. 5.3 Different presentations of a unit cell of high-temperature (*cubic*) form of perovskite structure (**a**)(**c**). With $BaTiO_3$ this form is transformed at about 120 °C in a tetragonal one, having a single $\langle 001 \rangle$ axis of symmetry and Ti^{4+} slightly displaced in the $\langle 001 \rangle$ direction (**b**)

such a degree that a spontaneous polarization, i.e. a Polarization in infinitely small EM fields, becomes possible (see Eq. (5.1).

Pyroelectrics The spontaneous Polarization can be easily modified by, respectively, heat (in pyroelectrics) and external near EM field (in ferroelectrics). If the structure has a single axis of symmetry, pyroelectric properties can occur. Figure 5.4 schematically shows rows of spontaneously formed dipoles extending in the $\langle 001 \rangle$ direction, normal to the surface of the material. If such material is in an electric circuit, between two conducting electrodes, the dipole charges on the materials surface compensate and bind some mobile charge carriers of an opposite sign in the electrodes (Fig. 5.4a). Let a material surface, e.g. one so polarised that positive dipole poles dominate, be subjected to a source of heat, thus becoming hot. Due to excitation of the movements of ions at an increased temperature, the original orientation of the dipoles is disturbed and fewer negative charge carriers) in the electrode (i.e. electrons) are bound by the positive dipole poles (Fig. 5.4b). The resulting increase in free electron concentration on this electrode causes an increase in voltage (in an open circuit) or current flow (in a closed circuit) between electrodes. Such effects are significant, only a slight increase in material surface temperature, for instance, of 0.1 °C, may cause voltage of up to 100 V. Modern measuring equipment enables measurement of voltage changes in fractions of volts, so it is easy to observe the appearance of weak sources of heat in the immediate vicinity or very intense but distant sources (thermal machinery, missile engines).

Ferroelectrics In some ionic dielectrics with one axis of symmetry and a specified Polarization direction, small domains (10^{-3}–10^{-4} m) with opposing Polarization directions (Fig. 5.5a) are formed, then initially the material is not polarised macroscopically, but after being subjected to an external EM field, shows properties typical for ferroelectrics. The external EM field can here easily change its direction of Polarization through the growth of domains with dipoles oriented according to

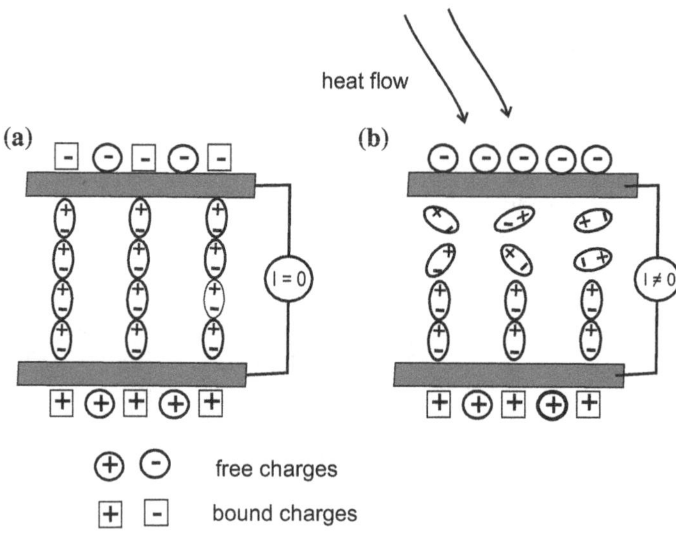

Fig. 5.4 Pyroelectric effect, explained in detail in the text

the field direction (Fig. 5.5b). This feature of easily switchable Polarization direction ensures a broad range of applications for ferroelectrics, including, but not limited to, computer memories and intelligent cards (→1. A brief history of ceramic innovations).

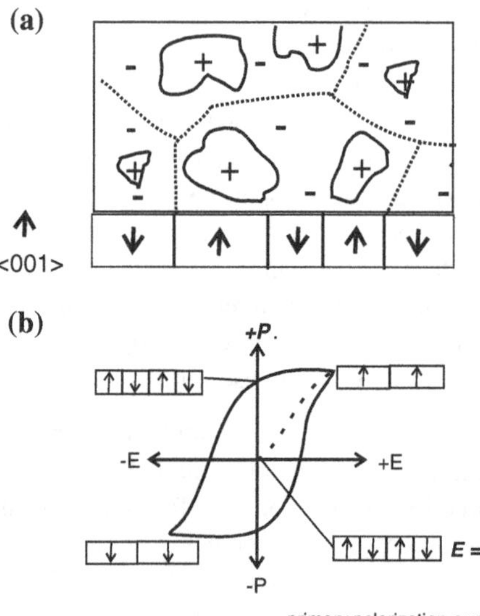

Fig. 5.5 Horizontal and vertical projection of ferroelectric domains in polycrystalline $BaTiO_3$: (**a**); Polarization changes in $BaTiO_3$ under influence of an external EM due to growth of domains in which the Polarization direction is identical or similar to the field direction (**b**)

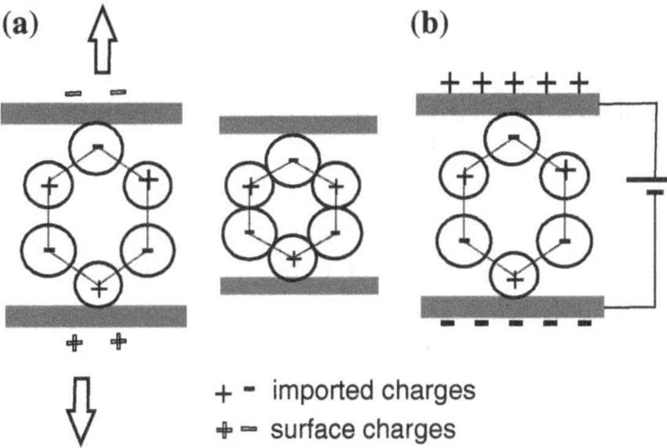

Fig. 5.6 Piezoelectric effects in a dielectric material. Direct piezoelectric effect: due to strain there is an outward shift either of cations or anions on opposite surfaces, which generates surface charge and potential difference between the surfaces (**a**). Reverse piezoelectric effect: surface charge caused by application of an external EM field attracts ions with an opposite charge in the material; expansion of the material is the result (**b**)

Piezoelectric Effects An example of dielectrics which cannot be considered as purely ionic dielectrics are αAl_2O_3, Si_3N_4 and SiO_2.... The bond-forming elements have here similar effective nuclear charges Z_{eff}. Due to this the bond-forming elements are expected to have a tendency to share their valence electrons and form bonds called by Pauling partly polarised covalent bonds (Fig. 5.2b). The covalent contribution to the bonds permitted to exploit αAl_2O_3 and Si_3N_4 as structural ceramics (→3. Ceramics to overcome brittleness) while as dielectrics these compounds have low electric susceptibility.

Nevertheless, even such e-materials retain the features of ionic bonds in a degree permitting to share piezoelectric effects and electrostriction with ionic dielectrics. If there is no centre of symmetry y in the structure of the dielectric, charges on its surfaces can be created by deforming it with a mechanical load, e.g. by expanding (Fig. 5.6a) or compressing it. A potential difference and voltage V between the opposing surfaces of the material is the result. This is called a direct piezoelectric effect. Conversely, the application of an external EM field creates on the opposing surfaces different charges attracting ions of opposite charge in the material. This brings about elongation of the material (Fig. 5.6b). This is called an inverse piezoelectric effect. Both piezoelectric effects are common in many applications, including force actuators of power up to 1,500 [$W \cdot cm^{-3}$], three orders of magnitude higher than hydraulic, pneumatic and EM actuators, and acoustic sound generators and sensors.

Electrostriction A behaviour related to the inverse piezoelectric effect is electrostriction, i.e. changes in volume and shape of the materials due to application of a near EM field. In contrast to piezoelectricity, electrostriction is a nonlinear effect which occurs with all dielectric but it is pronounced only in a group of dielectrics, called relaxers.

5.2 Unusual Conductors

The unusual conductors discussed are materials with very high-effective electrical conductivity at room temperature, like graphene and carbon nanotubes, composed of graphite-like layers (Figs. 5.7 and 5.8); as well as 'high-temperature' superconductors, like the cuprates—(Fig. 5.9). According to the effective nuclear charge of the bond-forming elements, the C–C and Cu–O bonds belong to partly polarised ones, like the case in many dielectrics. This indicates that, at least in the 'high-temperature

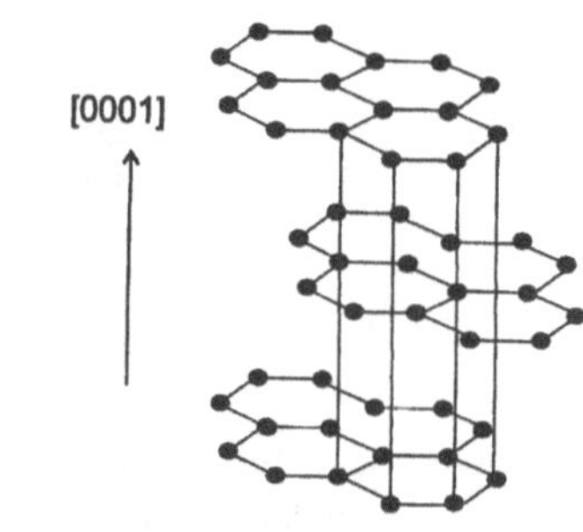

Fig. 5.7 Structure of (*hexagonal*) graphite

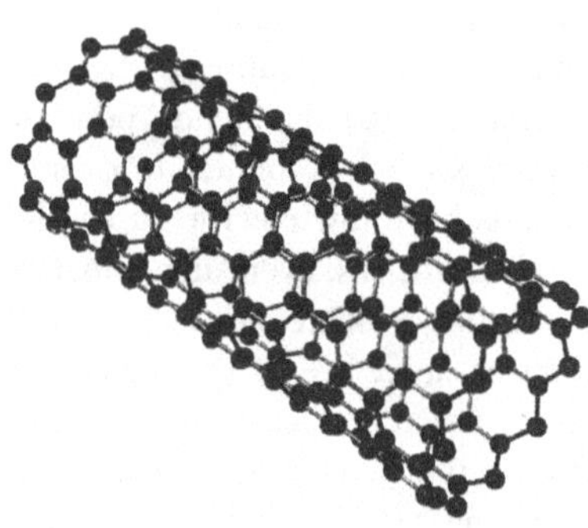

Fig. 5.8 Structure of a (*single-wall*) carbon nanotube

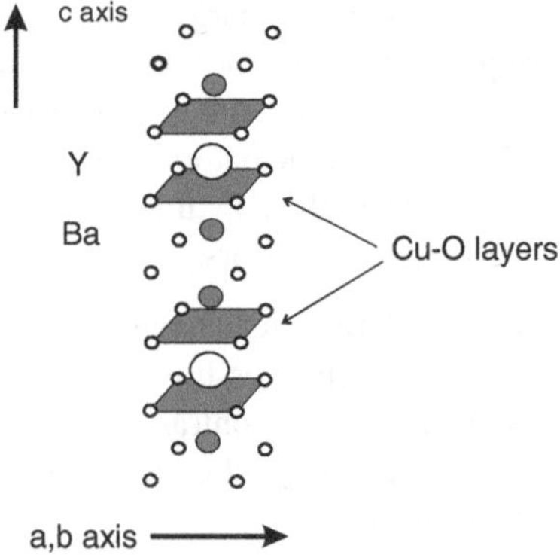

Fig. 5.9 Unit cell of $YBa_2Cu_3O_{7-x}$ structure

superconductors', the conduction is a different phenomenon from normal electric conductivity. However, the mechanism of the unusual conductivity is still far from understood. Therefore, we shall refrain from any further comment except mentioning that the common feature of the unusual conductors discussed here is a structure composed of weakly bound two-dimensional atomic layers with bonds having covalent contribution.

Structure and Properties of Graphite and Related Materials The electronic structure of carbon atoms is written as $1s^2 2s^2 2p^2$. This means that its four valence electrons are distributed in orbitals 2s and 2p of the layer L = 2. In the case of graphite, initiation of an electron from one orbital 2s to the empty third orbital 2p causes the formation of three orbitals 2p and hybridisation of two of them with the remaining orbital 2s. Three mixed hybridised sp^2 orbitals are created. Carbon atoms with such characteristics form $[CC_3]$ coordination groups in which three carbon ligands, located on one plane, are bonded by covalent bonds to a central carbon atom. The bonds of ligands and central atoms create a flat angle of 120° between them. By combining $[CC_3]$ coordination groupings, two-dimensional layers of carbon atoms are formed, called graphenes. The structure of graphite comprises a series of weakly bound graphenes As regards current conductivity, it is important that one of the orbitals 2p of carbon atoms is not involved in the creation of covalent bonds in graphenes. The electrons of this orbital are therefore practically free and cause the significant electric conductivity of graphite. It was found that in a near EM field, they move very easily from one position to another along graphene layers. This ease is reflected by the low-effective mass of those electrons and thus by the high mobility of the electrons.

This is illustrated in Table 5.1. According to this data, this mobility, already high for graphite, increases by one order of magnitude in carbon nanotubes, which can be described as formed by rolling single carbon layers of graphite structure into tubes (Fig. 5.9), and by two orders of magnitude in isolated single flat graphenes.

As materials with highest effective conductivity at temperatures close to room temperature, carbon nanotubes and graphenes have begun to be used in various devices such as MOSFET transistors (→1. A brief history of ceramic innovations).

Superconductivity is a material's ability to conduct an electric current without resistance and thus without loss. In contrast to superconducting metals, which have to be cooled to –230 °C to enable superconductivity, in such cuprates as $YBa_2Cu_3O_{7-y}$ (Fig. 5.9) and in other ceramic cuprates such as $HgBa_2Ca_2Cu_2O$ x or $Ba_2Sr_2Ca_2Cu_3O_{10}$, superconductivity was found at higher critical temperatures T_K

Table 5.1 Electron mobility in conductors

Material	Electron mobility b_e ($cm^2 \cdot V^{-1} \cdot s^{-1}$)
Inorganic semiconductor (heavily doped Si)	1,400
Organic semiconductors	<10
Carbon nanotubes	1,00,000
Graphene	2,00,000 (low temperature)

Table 5.2 Superconductors and their critical temperatures

Year of discovery	Material	T_K (K)	
1915	Pb	10	metals
1952	Nb_3Sn	14	
1970	Nb_3Ge	30	
2000	MgB	40	
1986	$YBa_2Cu_3O_{7-x}$	89	ceramics
1990	TaBaCaCuO	125	
1990	HgBaCaCuO	150	

than for the metals, from around −140 to −196 °C (Table 5.2). This explains the name 'high-temperature superconductors'.

Superconductivity of Cuprates Despite the practical use of this phenomenon (see, e.g. below), the phenomenon of superconductivity has not yet been explained. In this situation, it is better to focus only on the experimental facts. For example, it was found that the motion of electrons contributing to superconductivity in the above-mentioned cuprates is parallel to the Cu–O planes of their laminar structures (Fig. 5.9), and the best superconducting properties are achieved in the case of thin layers of those compounds in which there is a parallel orientation of Cu–O planes to the direction of electric charge transport.

Electroenergetic Transmission Lines with Superconductors The critical temperatures of the discussed ceramic high-temperature superconductors are lower than the boiling point of liquid nitrogen (77 K). Such temperatures are relatively easy to achieve given the current state of cryogenic engineering. Thanks to this, the first power transmission line has been built using cables with high-temperature superconductors as the current carrier. Power transmission systems with such cables ensure high transfer efficiency: with zero resistance, it is possible to increase current intensity by a factor of 150 compared to copper cables. It is also possible to reduce energy loss as a result of interference currents (short circuits). The short circuits are becoming more probable due to the increased density of power transmission systems nowadays. When a certain critical current intensity value is exceeded, superconductors revert to 'normal' condition and resistance appears, preventing overcurrent during short circuits.

Chapter 6
Materials Versus Light

6.1 Exchange of Energy with Remote Electromagnetic Fields

Ceramics are one of the most important types of materials in which valence electrons can exchange energy with remote electromagnetic fields, including the visible range of solar radiation.

According to the accepted conception of dual properties of light, the energy of light is transmitted by waves but its exchange with the electrons occurs by discrete portions, quanta, called photons. In the original Einstein model, the relation between the energy of photons (hv) and the wavelength of light is given as

$$hv = h\,c_o/\lambda \tag{6.1}$$

where: h—Planck's constant (h = 6.626176.10^{-34} [J.s]); v—wave frequency (the number of cycles of wave state repeating in a unit of time); λ—wavelength, i.e. the distance between neighbouring points for which the electromagnetic field of radiation is the same c_0—speed of light in a vacuum (c_0 = 2.997924.10^8 [m.s^{-1}]). Table 6.1 presents photon energy and wavelengths for different parts of the electromagnetic spectrum, including the visible range.

Light acting on a material can be reflected from the materials surface, penetrate it or be absorbed. According to the energy conservation principle, for any wavelength of light:

$$R(\lambda) + T(\lambda) + A(\lambda) = 1 \tag{6.2}$$

where: $R(\lambda)$, $T(\lambda)$ and $A(\lambda)$ denote, respectively, the ratio of the intensity of light which is, respectively, reflected from, passes through, or is absorbed by the material and the intensity of incident light; these are called, respectively, reflectance R, transmittance T and absorbance A. In terms of interaction with visible light, ceramic

R. Pampuch, *An Introduction to Ceramics*,
Lecture Notes in Chemistry 86, DOI 10.1007/978-3-319-10410-2_6

Table 6.1 Electromagnetic radiation spectrum

Radiation	Wavelength range λ(m)	Photon energy (eV)
Cosmic rays	10^{-16}–10^{-13}	>1.10^7
Gamma rays	10^{-15}–10^{-10}	>1.10^4
X-rays	1.10^{-13}–1.10^{-8}	>1.10^2
UV radiation	1.10^{-9}–$3.9{\cdot}10^{-7}$	>3,184
Visible range radiation	$3.9{\cdot}10^{-7}$–$7.7{\cdot}10^{-7}$	1,656–3,184
Infrared radiation	$7.7{\cdot}10^{-7}$–1.10^{-3}	<1
Microwaves	1.10^{-3}–1	1.10^{-2}–1.10^{-5}
Radiofrequency waves	1–1.10^4	1.10^{-6}–1.10^{-9}

materials can be divided into transparent, where $T(\lambda) \neq 0$ and non-transparent, where $T(\lambda) = 0$.

6.2 Non-transparent Materials

As indicated in the introduction, absorption of photons of the visible electromagnetic radiation results from their interaction with valence electrons, which are relatively easily excitable charged particles. Photons can be absorbed and transfer their energy to the electrons only when their energy $h\nu$, corresponds to the difference in energy of the ground level and the first excited level of the valence electrons, provided the higher level is not fully occupied (Fig. 6.1a). The energy of excited electrons is represented as

$$E_{\mathrm{m}} = E_{\mathrm{n}} + h\nu_{\mathrm{photon}} \tag{6.3}$$

where: E_{m}—higher electron energy level; E_{n}—lower electron energy level; $h\nu_{\mathrm{photon}}$—photon energy.

The energy of valence electrons can be conveniently illustrated by band structure models (Fig. 6.2). The band models show the ranges of allowed levels of valence electron energy projected onto a direction coordinate in the material. In ceramic materials the band of ground energy levels (the valence band) is separated from the band of energy levels of excited electrons (the conduction band) by an energy gap E_{g}. Photons can thus be absorbed if $E_{\mathrm{m}}-E_{\mathrm{n}} = E_{\mathrm{g}}$ does not exceed $h\nu$. The highest energy of visible light photons is about 3 eV (Table 6.1) and the occurrence or absence of their absorption is a criterion in the classification of ceramic materials into non-transparent semiconductors where $E_{\mathrm{g}} < 3$ eV and dielectric materials (→5. Unusual dielectrics and conductors) where $E_{\mathrm{g}} > 3$ eV.

Photovoltaic Effect Electron excitation by photons (Fig. 6.1a) in semiconductors is increasingly used for creating the photovoltaic effect, i.e. generating voltage. The photovoltaic cells, usually with silicon as the semi-conducting material, are

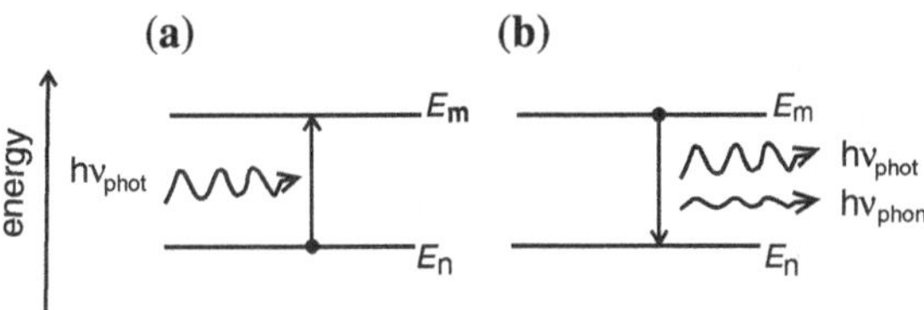

Fig. 6.1 Absorption of photons by exciting electrons to higher energy levels (**a**) and emission of photons and phonons on the return of excited electrons to the ground energy level (**b**)

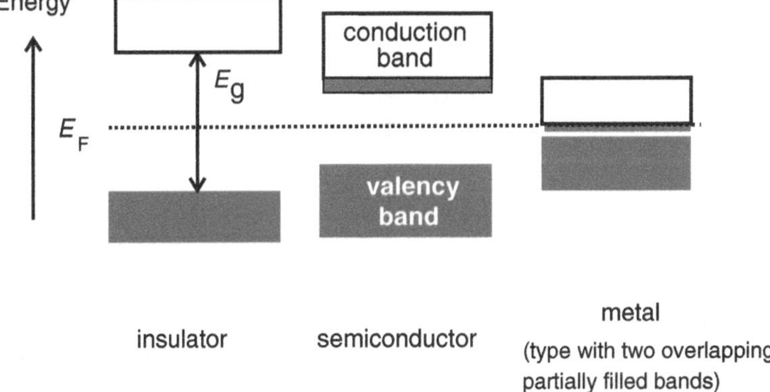

Fig. 6.2 Bandwidth models of solids. Electrons of atoms in solids interact and, when located in the same space, cannot have four identical quantum numbers (Pauli exclusion principle). Therefore, their energy splits into very close lying levels and energy bands are created. In the diagram, only the band formed by valence electrons (valence band) and the immediately higher band (conductivity band) are shown. In ceramic dielectrics and semiconductors these bands are separated by a band gap, E_g—an inadmissible energy range for the electrons. *Note* the occupation of bands by electrons is indicated by shadowing. iE_F denotes the Fermi level which equals the chemical potential of an electrons and, in a different approach, stands for the energy level at which the probability of electron occupation equals 50 %

accordingly called solar batteries. Photons acting on the material excite valence electrons (Fig. 6.1a) which leave electron holes in the valence band. To generate a potential difference and voltage, negatively charged electrons and positively charged electron holes have to be prevented from recombining immediately. This can occur if there is in the material a p–n junction. The p–n junction is an interface between extrinsic p-type semiconductor (with electron holes as the majority charge carriers) and n-type semiconductors (with electrons as the majority charge carriers). The p–n junction features a depleted zone from where the mobile charge carriers (electrons and electron holes) had diffused away, leaving uncovered space charge. The space charge brings about the desired drift of excited electrons and electron holes apart (Fig. 6.3).

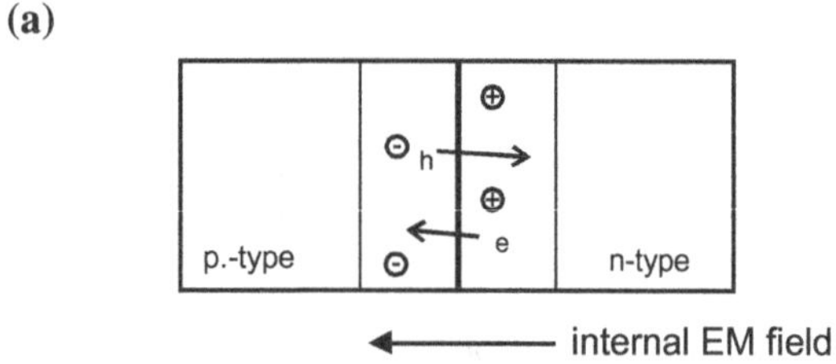

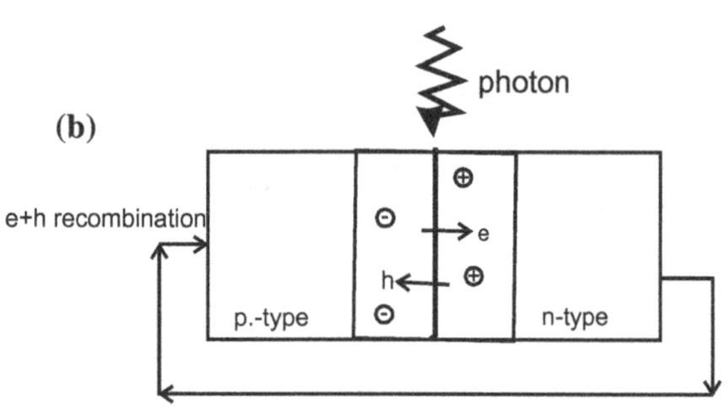

Fig. 6.3 The p–n junction: creation of a zone depleted in mobile electrons, e, and electron holes, h, in which also a space charge has been formed by uncovered acceptor '−' and donor ions '+' (**a**); separation of electrons excited by photons and of electron holes due to their drift caused by the space charge (**b**)

Luminescence Luminescence is the reverse process which involves emission of photons when excited valence electrons (in the conduction band) are recombined with electron holes (in the valence band) (Fig. 6.1b). The global energy lost by the electrons in this process is equal to the width of the band gap E_g. If all of this energy is used for the generation of a photon with energy hν, the wavelength of light emitted is given by Eq. (6.1):

$$\lambda = hc_0/E_g \tag{6.4}$$

where: λ—wavelength of light; h—Planck's constant; c_0—speed of light in a vacuum; E_g—band gap width.

Each wavelength range of the visible radiation gives a different colour impression. Since, there are different values of E_g in individual semiconductors, each type of a semiconductor thus emits light in different colours. All of this information is shown in Fig. 6.4 for semiconductors with intense luminescence, such as GaN, GaAs a.o. In some semiconductors, such as Si, Ge and SiC, energy lost by excited electrons, when they regain their ground state, generates both photons and phonons (Fig. 6.1b). The latter term refers to the energy quanta of correlated oscillations of atoms around their equilibrium positions (Fig. 6.5). Following the energy conservation rule, it can be stated that in this case:

$$E_m - E_n = h\nu_{\text{phot}} + h\nu_{\text{phon}} \tag{6.5}$$

where: E_m—higher electron energy level; E_n—lower (ground) electron energy level; $h\nu_{phot}$—photon energy; $h\nu_{phon}$—phonon energy. As the excited electrons

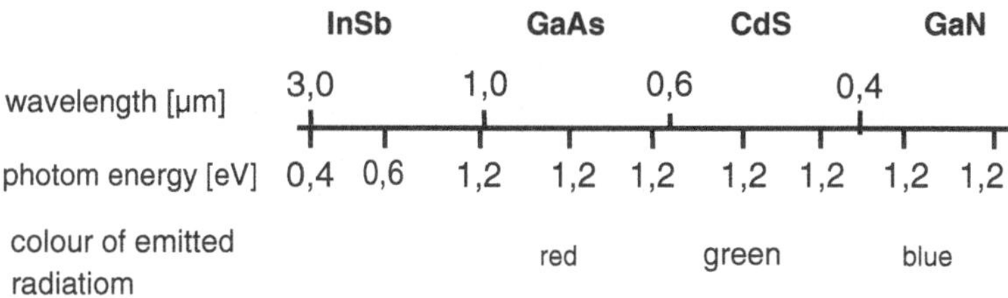

Fig. 6.4 Energy and wavelength of photons and colour of light emitted by some semiconductors

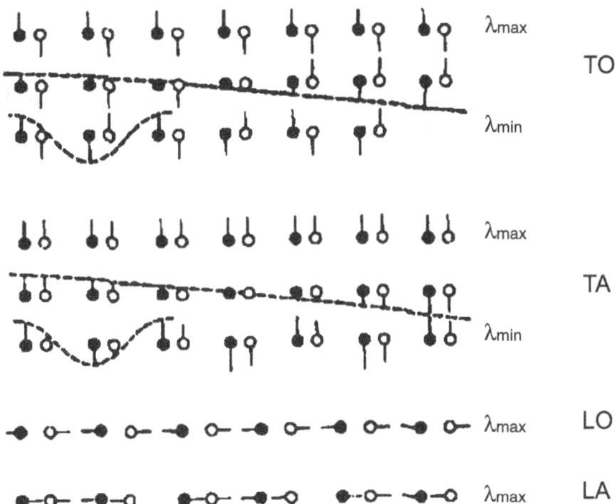

Fig. 6.5 Vibrations of a linear atomic structure. Note: *TO* transverse vibrations of then optical branch; *LO* longitudinal vibrations of the optical branch; *TA* transverse vibrations of the acoustic branch; *LA* longitudinal vibrations of the acoustic branch. The frequency and energy of vibrations normally increase in the sequence *LA* < *TA* < *LO* < *TO* at each wavelength

loose a considerable part of their energy to generate phonons, there is less energy left for the emission of photons.

It is difficult to explain in plain language why photons are predominantly generated in some cases and photons plus phonons in others. It is said that, in contrast to typical photon-emitting semiconductors (GaAs, GaN, Cds) there is in Si, Ge and 4H-SiC an indirect bad gap. This term is used to stress the point that in this case, the excited electrons (in the conductivity band) and the electron holes (in the valence band) have a different momentum. Returning to the basic state, the electrons have thus to pass through an intermediate state in which they transfer a portion of their energy to phonons, so that the momentum of electrons and electron holes becomes equal. It should be noted in this connection that Si and 4H-SiC have bonds with f a covalent contribution, while the bonds in GaAs, CdS, GaN are highly ionic (→5. Unusual dielectrics and conductors).

Electroluminescence Luminescence requires a constant and intense recombination of electrons with electron holes. This is achieved by using semiconductor diodes, where n-type semiconductor and p type semiconductor are connected by the p–n junction. With the diode forward-biased, i.e. when the polarity of applied voltage makes the p–n junction conductive, electrons of n-type semiconductor are injected into the conductivity band of p type semiconductor, and recombine with electron holes (Fig. 6.6). Since the semiconductors are non-transparent, emission of photons generated in this way is possible only from surface layers of the material. To achieve a high ratio of surface development to volume, light-emitting diodes have to be miniature. The efficiency of converting electric energy into light is here up to 100 %, at a voltage from several to a dozen or more volts. All this has led to the widespread use of light-emitting diodes (LEDs) in making lamps with a brightness many times greater than that of traditional bulbs. Their miniature size permits to apply LEDs for illumination of displays and keypads of cellular phones. Thanks to the use of LEDs, a revolution can be expected in the lighting of streets, houses and cars.

Electroluminescent diodes are not the only way of practical application of luminescence. Luminescence of materials subjected to action of various types of electromagnetic radiation is used for its detection, for road signs visible at dusk and at night, and many other applications.

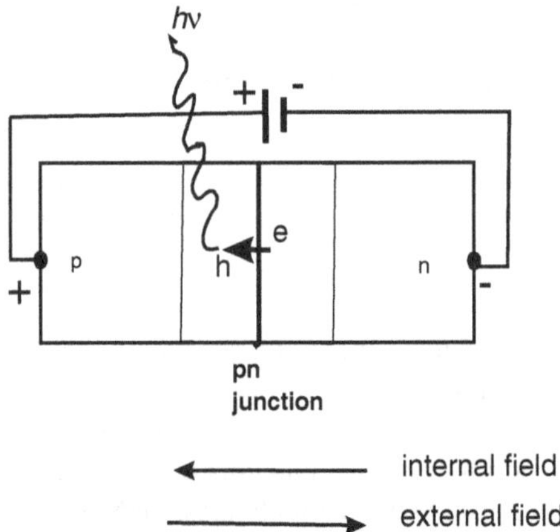

Fig. 6.6 A semiconductor light-emitting diode (LEDs). The p–n junction is forward-biased (by connecting positive voltage to a p type p semiconductor and negative voltage—to a n-type semiconductor) and the external electromagnetic field has a polarity opposite to the internal field in the semiconductor. Therefore, the resultant field is small and the p–n junction conducts electric current. This enables continuous injection of electrons (e) from the n-type semiconductor to the p-type semiconductor and their recombination with electron holes (h) resulting in an emission of photons of an energy hv

6.3 Transparent Materials

Let us consider the passage of a light beam from an optically rarer medium (denoted by index 1) to an optically denser one (index 2) and vice versa (Fig. 6.7). The ratio of the sines of the angle of incidence θ_i and of angle of the refraction θ_r is determined by the ratio of the phase velocities of light waves in both media (v_1/v_2). As this ratio corresponds to the ratio of refractive indices of the media (n_2/n_1) we get the Snell's empirical equation:

$$v_1/v_2 = n_2/n_1 = \sin\Theta_\iota/\sin\Theta_r \tag{6.6}$$

Oxide glass is the classic transparent material which has been applied for many years in the fabrication of lenses or prisms using the effects explained by the Snell equation the recent addition are optical glass fibres, in which the glass of fibre core has a higher refractive index (n_2) than that of the glass used as cladding (n_1). For a specified ratio n_2/n_1 and angle of incidence of the light beam on the core-cladding interface $\Theta_\iota = \arcsin(n_2/n_1)$, the angle of refraction Θ_r can attain the value $\pi/2$. Therefore, total internal reflection occurs and the light beam cannot escape from the optically denser core (Fig. 6.7b). Therefore, optical fibres can act as light tubes in which the light beam travels without leaving the fibre and are an important element of information technology (IT) networks (→ Fig. 1.13).

Maximum absorption of solar radiation is important for the efficiency of a number of devices, e.g. of photovoltaic cells. According to the principle of energy conservation (formula (6.2)) a high absorbance A can be obtained if a substrate of a low transmittance T is covered by a layer of a transparent material whose reflectance R at the air-material interface is low. The reflectance is given as

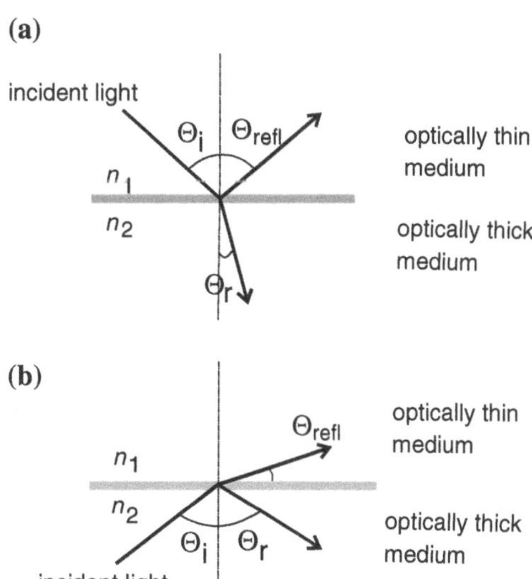

Fig. 6.7 Refraction and reflection at an interface of a beam of light entering an optically denser medium (**a**), an optically rarer medium (**b**)

$$R = ((n_1 - n_2)/(n_1 + n_2))^2 \quad (6.7)$$

Considering that n_1 is the refractive index of air ($n_1 = 1$), a refractive index of $n_2 \approx 1.5$ of the layer permits to attain a negligible reflectance, R. It should be noted in this context that refractive indices, absorbance, reflectance and transmittance depend upon the radiation wavelength. The wavelengths of the most intense sunlight reaching the earth range from $\lambda = 0.3$ to $\lambda = 0.9$ μm (Fig. 6.8) and it is in this wavelength range that the layers should have a refractive index of $n_2 \approx 1.5$. Such requirement is met by transparent layers of dielectrics, like the fluorides, αAl_2O_3, ZrO_2, and SiO_2.

Within a transparent material, colour centres can occur which are capable of absorbing photons of a narrow range of visible (white) light. Consequently, in light passing through the material (and/or reflected from it), an incomplete combination of wavelengths in that range occurs and the impression of colour is created (Fig. 6.9).

An important type of local colour centres are d-block and rare-earth cations with unfilled *d* or f orbitals (Fig. 6.10), hence, with unpaired electrons, because the energy needed to excite them to higher energy levels matches the photon energy. Thermally resistant ceramic compounds which contain such cations, like chromites (e.g. $MnCr_2O_4$) or ferrites (e.g. $MnFe_2O_4$)m are called cal ceramic pigments because they can sustain firing at high temperatures which property is needed for their application d in glazes (→ Chap. 2. The tradition continued. The silica tradition.). Particles of the pigments are dispersed in the transparent vitreous SiO_2-rich glaze (Fig. 6.11), creating an impression of colour. The glaze illustrated in the figure contains also dispersed particles of other phases, typically SnO_2 or $ZrSiO_4$, performing other functions than the pigments. The particles of latter compounds are optically denser media than the vitreous matrix and there is light refraction at the particle–matrix interfaces. Owing to the chaotic orientation of the interfaces, the refraction occurs in different directions, and diffuse reflection predominates.

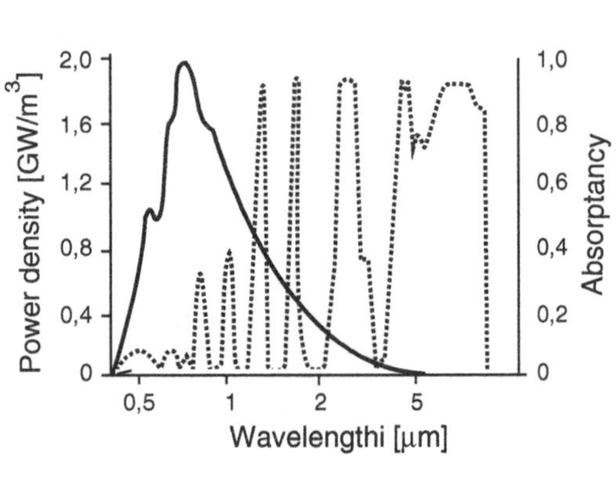

Fig. 6.8 The solar radiation spectrum and its absorption in the earth's atmosphere

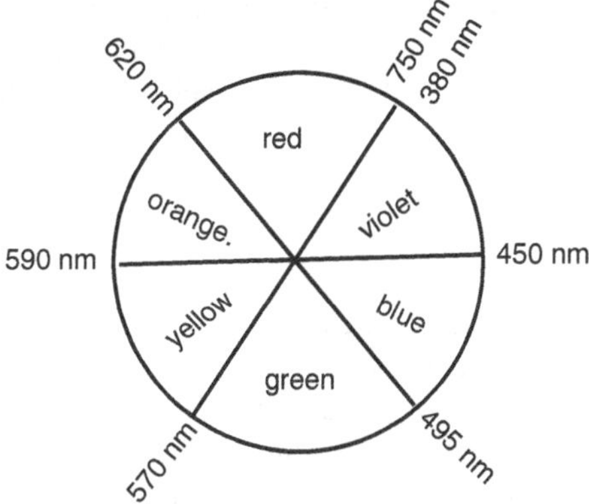

Fig. 6.9 Colour circle used to determine the perception of material colour resulting from absorption of a part of visible light spectrum by material. For example: absorption of light with a wavelength corresponding to red (620–750 nm) causes an impression of green and vice versa, absorption of blue light causes an impression of orange and vice versa

	I	IV	V	VI	VII	VIII	VIII	VIII
4		^{22}Ti	^{23}V	^{24}Cr	^{25}Mn	^{26}Fe	^{27}Co	^{28}Ni
	^{29}Cu							
5		^{40}Zr	^{41}Nb	^{42}Mo	^{43}Tc	^{44}Ru	^{45}Rh	^{46}Pd
6		^{72}Hf	^{73}Ta	^{74}W	^{75}Re	^{76}Os	^{77}Ir	^{78}Pt
	^{79}Au							

Fig. 6.10 Elements in the periodic table whose cations form colour centres in ceramic pigments

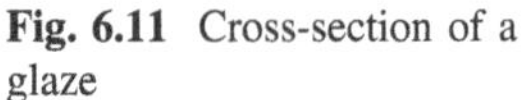

Fig. 6.11 Cross-section of a glaze

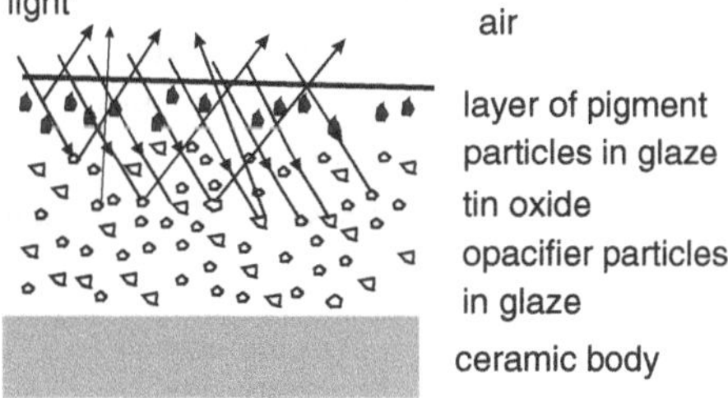

Moreover, at an appropriate concentration of the particles all incident light beam is reflected which makes the glaze opaque. The particles are most effective when their sizes is comparable to the wavelength of visible light, e.g. ranging from 100 to 1,000 nm. In absence of ceramic pigments a glaze containing dispersed particles of this type, called opacifiers, appears as an opaque, white object. Such phenomena are relevant also for polycrystals of inherently transparent materials. Interfaces between phases of a different optical density may occur in them, and light is reflected at the interfaces in the way discussed before. These include: gas-solid interfaces formed by pores pore-and and cracks, and grain boundaries. A high concentration of such interfaces can render the material opaque, again when the interfaces have a size comparable to the wavelength of visible light.

Luminescenceand Laseres Some transparent materials are used in generation of special luminescence, in which is created a concentrated beam of high-intensity light with a very narrow frequency (wavelength) range. This type of generation occurs in lasers. Only one type of laser will be described here, in which the luminescence is due to excited electrons of *d* orbitals of cations of rare-earth or d-block elements. In this context, the cations are referred to as luminophores which are incorporated into a transparent matrix. Typical luminophore–transparent matrix pairs are presented in Table 6.2. The electrons are excited by an external source of a high energy, inducing photons in the way shown in Fig. 6.1b. A photon, passing next to the cation still with an excited electron, causes a disturbance of this state and stimulates emission of another photon with the same energy $h\nu$ as in the passing photon. Since, the stimulating photon does not lose energy, two photons with identical energy $h\nu$, frequency and phase are created (Fig. 6.12b). Photons are forced to remain in the matrix for some time by reflecting them from semi-transparent mirrors. Therefore, the process of stimulated electron emission is repeated many times. This leads to a rapidly growing number of photons with the same energy $h\nu$ and continuing until the beam of light achieves a high-intensity limit which lets it escape from the matrix.

Table 6.2 Composition of oscillation rods and cations acting as luminophores in solid-state lasers

Cation	Rod	Wavelength of emitted radiation (μm)
Ti^{3+}	Al_2O_3, AB_2O_4 spinel[a]	0.65–1.1
Cr^{3+}	Al_2O_3, $LiYF_4$, $Ba_8Y_2F_8$	0.8–0.9
Cr^{4+}	$Y_3Al_5O_{12}$ (YAG), $MgSiO_4$	1.1–1.65
Nd^{3+}	Silica and fluorine glass, YAG	1.1

[a] *A* cations at tetrahedral positions, *B* cations at octahedral positions in spinel structure

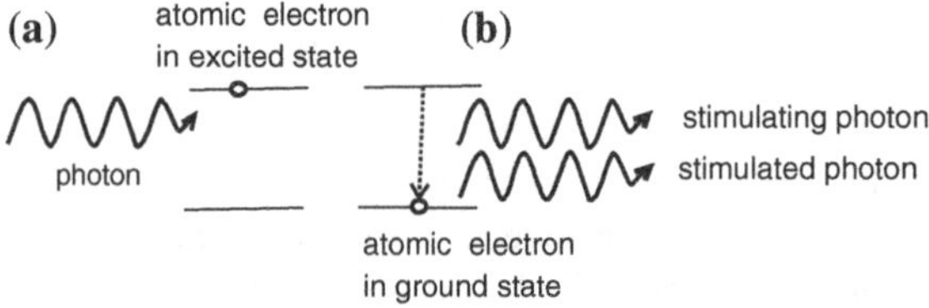

Fig. 6.12 Stimulated emission of photons. Explained in the text

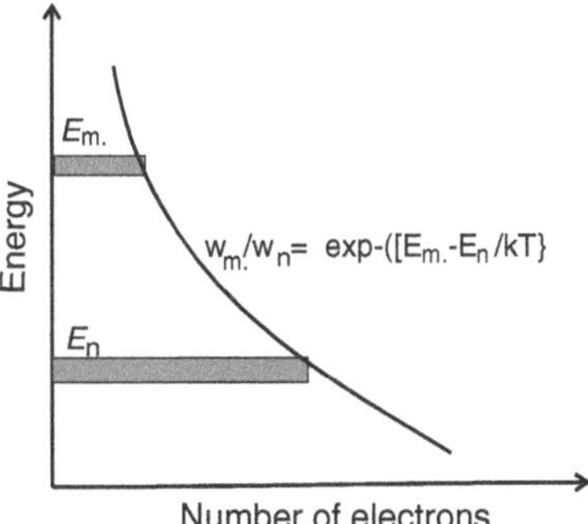

Fig. 6.13 Population of energy levels energy by electrons in equilibrium. *Note* the curve shows the ration of population of the higher (w_m) to the one of the lower energy level (w_n), corresponding to a Boltzmann distribution

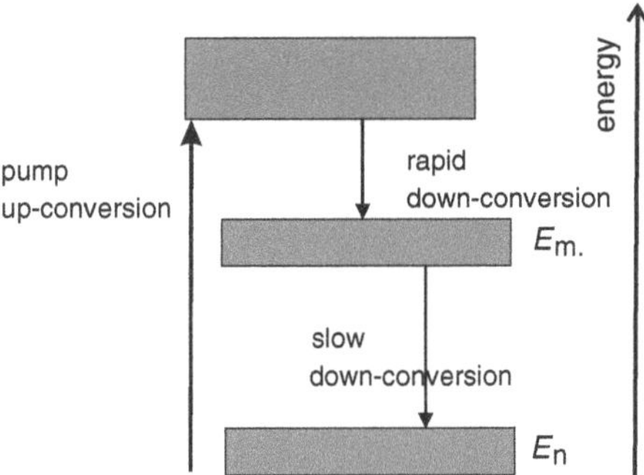

Fig. 6.14 Three-level model of luminescence. An external source of energy excites (pumps) electrons to a state of very high energy; rapid transitions of electrons to intermediate metastable energy states are followed by slow light-emitting (optical) transitions back to the ground state

The properties of the matrix material are important for the effectiveness of these phenomena. The matrix not only has to be transparent to enable photons to travel in it, but also must ensure minimal interaction of excited electrons with phonons. Another condition for continuous light emission in lasers is also the population inversion, a state in which the number of electrons excited to higher energy levels is continually greater than the number of electrons which spontaneously recombine with electron holes at a given time. Because this is not the case in thermal equilibrium (Fig. 6.13), the population inversion can be achieved first by exciting the electrons into very high energy states, using an external source. Secondly, by remaining in the excited electrons for some time in metastable energy states before returning to their ground level (Fig. 6.14). It is again important that the energy of electrons is not dispersed through interaction with phonons. All requirements mentioned before are met by the matrix materials listed in Table 6.2.

Chapter 7
Imitating and Supporting Nature

7.1 Smart Materials and Intelligent Systems

Structures of Nature's Creations and Their Imitations Until recently, the development of materials was based on mechanics which, as an accurate and effective field of science and technology, was the discipline of choice throughout the Industrial Revolution. This approach, however, has some limitations. In many respects, materials made in this way are inferior to the creations of nature which have developed over the course of evolution to ensure optimum functioning and survival. One example might be the structure of human and animal bones and wood, which ensures high flexibility when subjected to load, enables the change in internal stress trajectories so as to bypass the weakest or damaged points. Thus, for example, wood has an ultimate strength equal to 1/10 of its chemical bond strength, which is difficult to achieve in synthetic materials. It is hence not surprising that in the middle of the twentieth century, nature's structures began to be imitated in some materials, such as fibre-reinforced composites, cellular or foam products (→3. Ceramics to overcome brittleness. →4. Voids are important).

Materials with 'Smart-Like' Behaviour The world of technology also became attracted by some materials, called 'smart' or 'intelligent'. These terms are used here too, but it is worth noting that they are quite exaggerated. As far as individual materials are concerned, only a very few demonstrate behaviour which could be classified as 'smart', for example, polycrystalline barium titanate. Its rare property among dielectrics and semiconductors is a change of resistivity over a narrow temperature range by several orders of magnitude (Fig. 7.1). Therefore, a $BaTiO_3$, resistance heating element connected as a load to an electrical circuit is able to automatically adjust its temperature to a specified level. Part of the energy of the current flowing through the element is dispersed as Joule heating. Consequently, the temperature of the heating element increases, but these result in an increase in resistivity R of the element. As predicted by Ohm's law ($I = V/R$), a reduction of current intensity I flowing through the heating element is the result. From this

R. Pampuch, *An Introduction to Ceramics*,
Lecture Notes in Chemistry 86, DOI 10.1007/978-3-319-10410-2_7

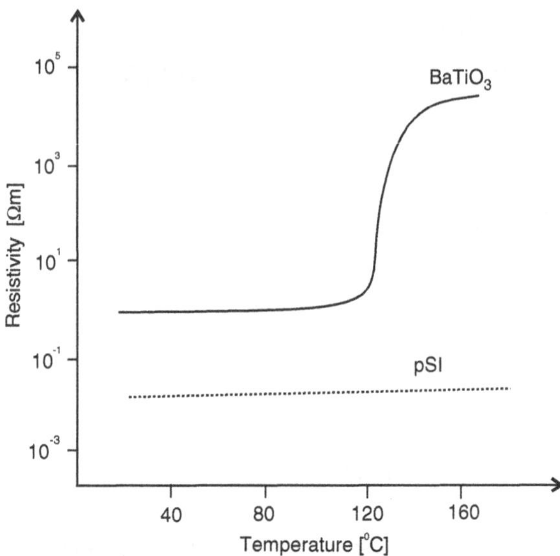

Fig. 7.1 Resistivity versus temperature for $BaTiO_3$

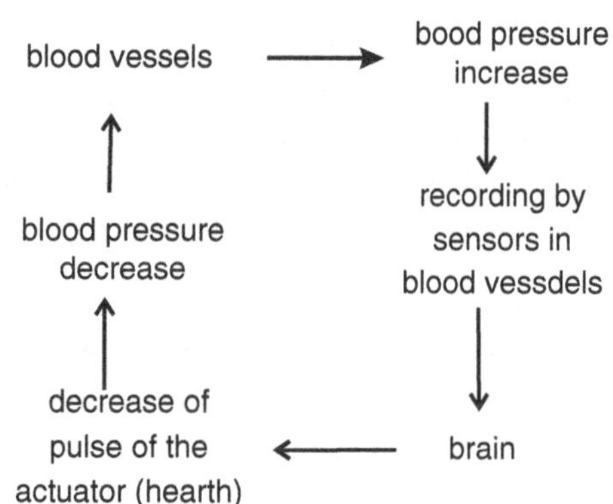

Fig. 7.2 Homeostasis

moment on heat losses due to radiation into the environment begin to outweigh the Joule's heat produced in the heating element, decreasing its temperature. Owing to this a decrease in resistivity *R* of the element occurs, due to which the current intensity and Joule's heating can increase again and the heating element regains its initial maximum temperature.

The example of $BaTiO_3$ shows that individual materials which could demonstrate 'smart' behaviour have to exhibit highly nonlinear responses to external stimuli. Besides barium titanate, such a response occurs in metallic shape memory alloys, and in thin layers of a few ceramic compounds such as $BiCeO_3$, $TbMnO_3$ or $Pb(Fe_{1/3}Ni_{2/3})O_3$.

The Principle of a Negative Feedback and Active Intelligent Systems However, nature functions intelligently. Namely, nature's creations are not only able to

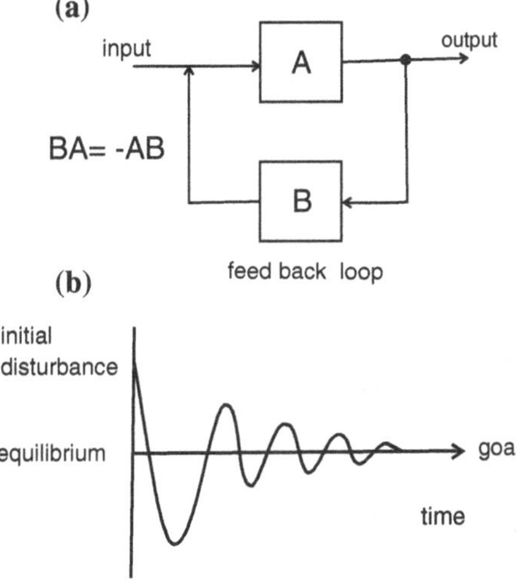

Fig. 7.3 Principle of (negative) feedback (**a**) and attenuation of deviation from equilibrium in time as a feasible result of negative feedback (**b**). *Note* more detailed explanation in the text

identify any changes in the environment disturbing their proper functioning but also to respond rationally and in a diversified way, proper for the situation. This ability can be described as being based on the principle of (negative) feedback, an example of which is homeostasis in adjusting blood pressure in living organisms (Fig. 7.2). Actions imitating homeostasis can be performed in so-called active intelligent systems (Fig. 7.3). One example of such systems is one where an electronic pulse (AB) indicating disturbances in controlled device A enters control device B via a feedback loop. The input of signal (AB) causes the output of signal (BA) from the control device, which has a sign opposite to that of the input and which enters through the feedback loop into device A (Fig. 7.3a). This and the following steps make it possible to either restore balance in A (Fig. 7.3b) or adjust the functioning of A to the changed ambient conditions. Typically, active intelligent systems comprise a sensor and actuator located, along with a control device, in the feedback loop path (Figs. 7.3a and 7.4). So far, mainly electrical phenomena have been exploited in the loop. Thus, a typical sensor is a converter which records chemical or physical stimuli from the environment (temperature, pressure, velocity, liquid flow, the presence of specific chemical molecules, etc.) and then, by changing voltage, impedance or capacitance, sends an appropriate electrical signal to the control device of the system. The control device is usually a microprocessor which functions similarly to a central processing unit (CPU) in a computer, sorting and classifying the input information received from the sensor and then analysing the situation based on the instructions stored in its memory. The result of the analysis is sent as an electric signal to the actuator, which converts it to the required physical or chemical action.

Active intelligent systems are currently used in diverse applications, such as monitoring and adjustment of combustion in engines, measuring blood pressure, eliminating vibration in cameras and adjusting the angle of attack of helicopter rotor

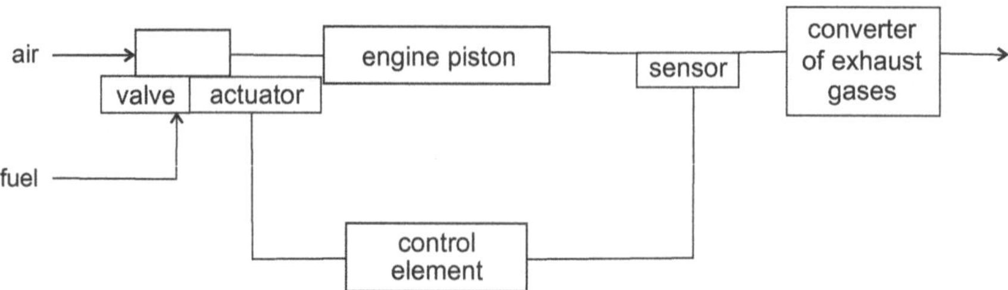

Fig. 7.4 An intelligent active system to monitor and adjust combustion in engines. A sensor (electrochemical cell) monitors the partial pressure of oxygen, its output voltage depends on the ratio of partial pressure of oxygen in air ($p_{O2\ (ref)}$) to partial pressure of oxygen in the combustion gas ($p_{O2(exhaust}$ An increase of $p_{O2(exhaust}$ at incomplete combustion is indicated by the sensor with increased voltage V. In response, the control component sends a voltage signal to the actuator. Thanks to the reverse piezoelectric effect (see Fig. 5.6b) the actuator expands and sets ajar the valves supplying fuel to the engine

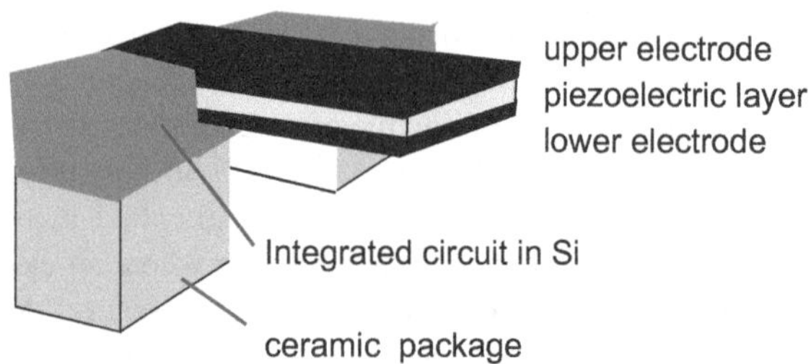

Fig. 7.5 Schema of a micro-electromechanic system (MEMS)

blades to pressure variations a.s.o. Figure 7.4 illustrates an active intelligent system to monitor and adjust combustion in car engines.

Micro-Electromechanical Systems The measure of the 'intelligence' of this and other systems is both the range and time of response to changes in the monitored equipment or environment. This is most often ensured by micro-electromechanical systems (MEMS), miniature components which measure and process such parameters as acceleration, pressure, distance, temperature, light and the chemical composition of an atmosphere. The heart of MEMS is a processing unit (microprocessor) made in the form of an integrated circuit on a silicon board. The board is then chemically etched and layers of other materials are deposited to make a sensor. Typical features of a MEMS fabricated in this way are shown in Fig. 7.5. Thanks to their small size, sensors in MEMS are characterised by low inertia, which makes it possible to quickly start and stop their functions. Moreover, thanks to their integration with the microprocessor in one substrate, their response to external stimuli is sent to the microprocessor almost immediately.

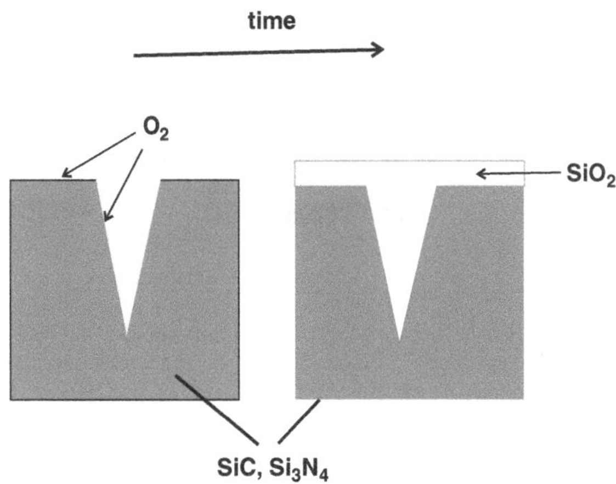

Fig. 7.6 Self-healing of cracks at an increased temperature (1,200–1,300 °C); in composites containing Si_3N_4 and/or SiC

7.2 Self-Healing Materials

Signs of the intelligence of living organisms include the processes of healing wounds. Thanks to the 'self-healing' of cracks, scratches and other defects, resembling these natural processes, materials can still function despite previous damage. In other words, the reliability of these materials can be enhanced. It has been possible to observe the self-healing of cracks and scratches a.o. in composites containing SiC or Si_3N_4. At increased temperatures and typical partial pressures of oxygen in atmosphere, oxygen atoms are incorporated into the structure of SiC or Si_3N_4, with formation of SiO_2 (solid up to 1,720 °C). This results in an increase in material volume, which, in turn promotes the self-healing of surface cracks (Fig. 7.6). The process is assisted by diffusion in SiO_2 towards the cracks under the force of chemical potential gradient between concave and convex surfaces (→Fig. 2.2). Another example comes from traditional ceramics. Namely, the self-healing of cracks in the lining of rubbish disposal sites, where it is necessary to prevent the penetration of hazardous waste into soil. The lining which behaves in this way comprises an internal layer of CaO- containing material A, and an external layer made of reactive pozzolanic material B (→2. The tradition continued. Lime tradition), e.g. fine SiO_2 from volatile ash (Fig. 7.7a). Should a gap occur in the lining, a highly exothermic reaction takes place in layer A, between CaO and soil moisture: $CaO + H_2O = Ca(OH)_2$. The reaction increases the local temperature to around 100 °C, which enables a reaction of the slaked lime, $Ca(OH)_2$ with the pozzolanic material of layer B. In this reaction calcium silicates are created which fill the gap (Fig. 7.7b). The number of ways to produce the self-healing of defects has been constantly growing.

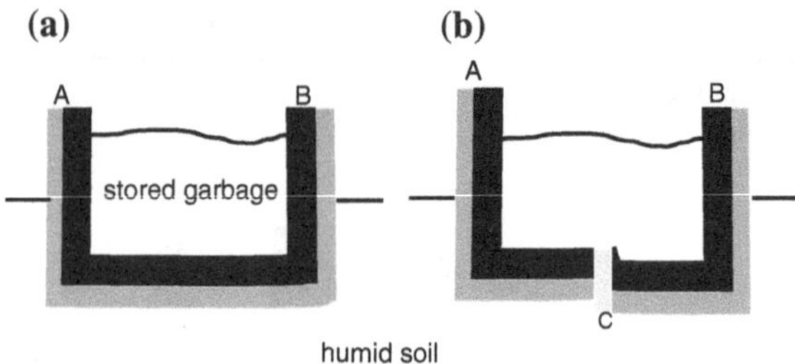

Fig. 7.7 Self-healing of cracks in sealing material of rubbish disposal sites. *Note* A—pozzolanic material, e.g. fine SiO_2 of volatile ash; B—Ca-containing material; C—material filling the crack due to a cycle of reactions: 1. $CaO + H_2O = Ca(OH)_2$; 2. $Ca(OH)_2$ + pozzolanic material → calcium silicate

7.3 Ceramic Biomaterials and Tissue Engineering

Properties of Ceramic Biomaterials Supporting nature through treating, strengthening or replacing tissues and entire organs is a practice initiated by the application of biomaterials (including ceramics) as implants in humans or animals. To be suitable for use, the biomaterials should have a number of properties. First of all, they must be tolerated by the body. Any foreign body (such as an implant) introduced to an organism causes an immune response whose intensity depends on the type of material. The strongest reaction is one in which the organism tries to isolate the foreign body from its tissues by means of an inflammatory reaction around it, which, after some time, results in the formation of granulation tissue. In a less drastic case, the foreign body may cause a fibroelastic response, which manifests itself as the creation of fibrous connective tissue encapsulating the implanted biomaterial. The thinner the shell, the greater the tolerance of the body to the implant.

More than half of the material of bones and teeth constitutes hard tissue of a composition corresponding to calcium hydroxyapatite $Ca_{10}(PO_4)_6(OH)_2$, which also contains CO_3, Na, Mg, K, Sr, Ba and Pb. The greatest tolerance in contact with this hard tissue is thus achieved by ceramic biomaterials with a composition similar to that of the biogenic calcium hydroxyapatite, or which contain calcium, e.g. silica glass with a calcium content higher than normal (bioglass). As a result of the complex reaction of dissolution-precipitation, implants made of these materials combine with the hard tissue of teeth or bones without a fibroplastic response. Liberation of small amounts of soluble calcium and silica by the implant also activates genes which create new and normal bone tissue. Similar processes cause bone tissue and synthetic hydroxyapatite to merge. Therefore, it became possible to use granulated synthetic hydroxyapatite to fill gaps in jawbones or limbs, and bioglass for solid middle ear implants.

Applications of Ceramic Biomaterials in Regenerative Medicine The constatation that some ceramic materials are biocompatible and promote growth of normal tissues, has fostered their use in regenerative medicine. The objective of regenerative medicine is the reconstruction, sustenance of life, and improvement in the functions of live cells in bones, blood vessels, skin, muscles or cartilage. Tissue engineering, an engineering branch of regenerative medicine, includes the following stages: 1. taking a section of live tissue; 2. using the section cells for creation of more complex structures (a stage called 2D cell expression); 3. growth of cells (e.g. osteoblasts) in three dimensions on the surface of a suitable matrix (scaffold); 4. implantation of a graft created in this way into the body.

Ceramics are used in stages 3 and 4 as the matrix (scaffolding) material. The greatest possible surface area is important here, since the tissues are formed on the surface of the matrix (scaffolding). This can be ensured by the use of highly porous matrix material composed of communicating pores. The combination of this requirement with good mechanical strength is ensured in particular by foam materials (→4. Voids are important). Another desirable property is degradation of the scaffolding material after integration of the implant within the body, followed by its elimination from the body.

Further Readings

F. Aldinger, V.A. Weberruss, *Advanced Ceramics and Future Materials* (Wiley-VCh, Weinheim, 2011)

R.S. Berns, *Principles of Color Technology*, 3rd edn. (Wiley, New York, 2001)

P. Boch, J.-C. Niepce, A. Bouquillon, *Ceramic Materials: Processes, Properties and Applications* (Wiley, New York, 2010)

R.J. Brook (ed.) *Concise Encyclopedia of Advanced Ceramic Materials* (The MIT Press, Cambridge, 1999)

H. Bruus, *Theoretical Microfluidics* (Oxford University Press, USA, 2008)

A.J. Burggraaf, I. Cot, *Fundamentals of Membrane Science and Technology* (Elsevier, Amsterdam, 1996)

C.B. Carter, M.G. Norton, *Ceramic Materials*, 2nd edn. (Springer-Verlag, Berlin, 2013)

F.A.L. Dullen, *Porous Media: Fluid Transport and Pore Structure* (Academic Press, New York, 1996)

G. Gautschi, *Piezoelectric Sensorics* (Springer-Verlag, Berlin, 2002)

L. Gibson, M.F. Ashby, *Cellular Solids:* (Cambridge University Press, Cambridge, 1999)

E. Görlich, *Effective Nuclear Charge and Electronegativity* (Polish Academy of Learning, Kraków, 1997)

M. Jouenne, *Céramique Générale* (Notions de Physico-chimie, Gautier -Villars, Paris, 1959)

A.A. Kelly, H. Macmillan, *Strong solids* (Clarendon Press, Oxford, 1986)

W.D. Kingery, H.K. Bowen, D.R. Uhlmann, *Introduction to Ceramics*, 2nd edn. (Interscience-J. Wiley, New York, 1976)

C. Kittel, *Introduction to Solid State Physics*, 8th ed. (Wiley, New York, 2004)

R. Pampuch, *Ceramic Materials: An Introduction to their Properties* (PWN-Elsevier, Warsaw-Amsterdam, 1976)

R. Pampuch, *ABC of Contemporary Ceramic Materials* (Techna Group, Faneza, 2008)

A. Petzold, *Anorganisch-nichtmetallische Werkstoffe* (VEB Deutscher Verlag für Grundstoff Industrie, Leipzig, 1981)

K.M. Rabe, J-M.Truscone, Ch.H. Ahl, *Physics of Ferroelectrics. A Modern Perspective* (Springer-Verrlag, Berlin, 2007)

J.S. Reed, *Introduction to the Principles of Ceramic Processing* (Wiley, New York, 1988)

Ch. Schacht, *Refractories Handbook* CRC Press (Taylor and Francis Publisher, Group), London, 2004)

H. Scholze, *Glass- Nature, structure and Properties* (Springer-Verlag, Berlin, 1991)

H.F.W. Taylor, *Cement Chemistry*, 2nd ed. (Thomas Telford Editions, London, 1997)

W. Vogel, *Glass chemistry*, 2nd edn. (Springer-Verlag, Berloin, 1994)

R. Pampuch, *An Introduction to Ceramics*,
Lecture Notes in Chemistry 86, DOI 10.1007/978-3-319-10410-2

Index

R. Pampuch, *An Introduction to Ceramics*,
Lecture Notes in Chemistry 86, DOI 10.1007/978-3-319-10410-2

Zeitfracht Medien GmbH
Ferdinand-Jühlke-Straße 7
99095 Erfurt, Deutschland
produktsicherheit@kolibri360.de